Gemini 7

The NASA Mission Reports

Compiled from the archives & edited
by Robert Godwin

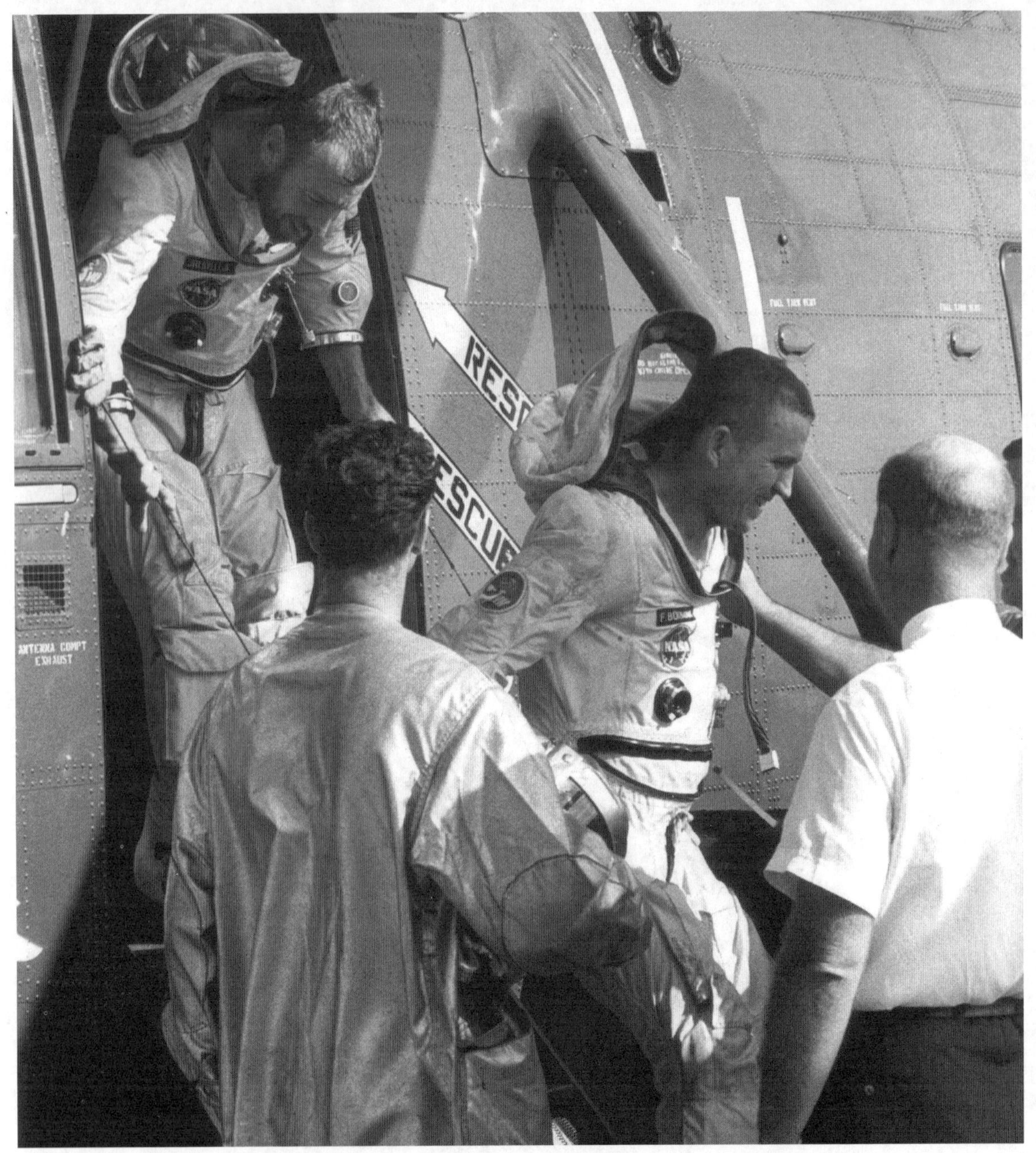

We acknowledge the financial support of the Government of Canada through the Book Publishing Industry Development Program for our publishing activities.
Published by Collector's Guide Publishing Inc., Box 62034,
Burlington, Ontario, Canada, L7R 4K2
Printed and bound in Canada
GEMINI 7 — THE NASA MISSION REPORTS
by Robert Godwin
ISBN 1-896522-82-3
ISSN 1496-6921
Apogee Books Space Series

GEMINI 7

The NASA Mission Reports

(from the archives of the National Aeronautics and Space Administration)

CONTENTS

FACT SHEET

EDUCATIONAL BRIEF

PRESS KIT

WORLD-WIDE COMMUNICATIONS NETWORK

MISSION OPERATION REPORT

LIST OF TABLES

PROJECT GEMINI DRAWINGS AND TECHNICAL DIAGRAMS

TECHNICAL DEBRIEFING

GEMINI VII / GEMINI VI LONG DURATION / RENDEZVOUS MISSIONS

GEMINI VII VOICE COMMUNICATIONS

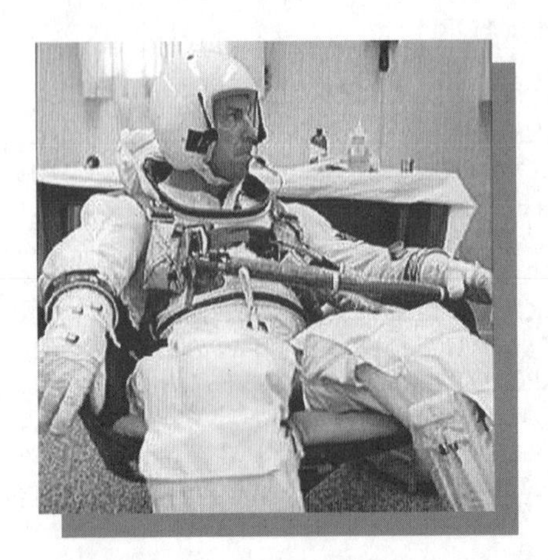

INTRODUCTION

Frank Borman and Jim Lovell spent two weeks together in a space the size of the front seats of a Volkswagen Beetle. It is difficult to fully comprehend the mindset of an individual that could live under such conditions. As if the close-quarters were not sufficient cause for alarm, the crew were obliged to conduct an impressive array of tests upon themselves. The goal was to study the long-term effects of space flight on the human body. The mission duration was two weeks because it was anticipated that it would take about fourteen days to get to the moon and back.

One of the few concessions made to comfort was the use of a new "soft" spacesuit which Lovell had helped design. The bizarre looking suit and helmet made the two astronauts look like they had just landed out of *Planet of the Apes* or some other science fiction film.

Once Gemini 7 launched into Earth orbit the ground crews scrambled to launch Gemini 6 to effect a rendezvous. While Borman and Lovell went about their routines, launch pad 19 was quickly prepared for another Titan-Gemini stack. After an almost catastrophic launch failure involving an abort, Wally Schirra and Tom Stafford ascended and flew to within about a foot of their companions in Gemini 7. It was an important accomplishment in the goal of getting to the moon.

Once Schirra and Stafford returned to Earth, Borman and Lovell continued to perform their studies of the human body. Fortunately the spacecraft systems operated almost flawlessly and the two men were able to work without interruption.

Some of the more unpleasant experiments required that they keep track of their calorie and water intake. This involved storing everything from urine to sweat and feces so that it could be studied back on Earth. Every single ounce of water was logged and recorded for two weeks, a chore that they made clear was unacceptable. Their meals were crumbling as they were opened and particles of debris cluttered the air around them, often getting into their eyes.

Meanwhile experiments were packed into the timeline, including the almost astonishing feat of actually monitoring the first launch of a submarine based missile. The Polaris was launched just as they passed overhead, an impressive logistical feat.

Another early attempt at cutting edge communications involved using a hand-held laser to send information back to the ground. In the world of today we use lasers for communications almost every time we pick up a telephone, but in 1965 it was a virgin technology.

Jim Lovell also had to use a sextant for navigation. The new star-sextant became standard operating equipment three years later when he and Borman used it to successfully navigate their way to the moon and back aboard Apollo 8.

After two weeks in zero-g, the two tired and bearded astronauts maneuvered their way home and were amazed at just how tough reentry proved to be. Commenting that the g-force seemed enormous after so much time in space.

It is interesting to note just how important their debriefing comments would have been back in 1965. Much of their mission involved working with the military. The debrief document would have been considered a treasure trove of information to the Soviet Union. Today it is a window into an era which seems distant and yet recognisable.

Long-term spaceflight was just an elusive goal until Gemini 7. Many scientists did not know if the crew would even be able to walk onto the recovery ship, but the tenacious and indomitable spirit of men like Borman and Lovell proved it could be done. The flight of Gemini 7 ironed out many details and pushed NASA an important step closer to winning the moon race.

Robert Godwin
(Editor)

NATIONAL AERONAUTICS AND SPACE ADMINISTRATION
WASHINGTON, D.C. 20546

FACT SHEET
THE GEMINI "7-6" MISSION

Gemini 7 Crew: Astronauts Frank Borman, James A. Lovell, Jr.
Gemini 6 Crew: Astronauts Walter M. Schirra, Jr., Thomas P. Stafford

On December 4, 1965 the United States of America launched the Gemini 7 spacecraft from Cape Kennedy, Florida, on a 14-day mission which was also to include a rendezvous with the Gemini 6 spacecraft to be launched December 15, with Gemini 7 as the target vehicle.

Both missions proceeded as planned with the two spacecraft accomplishing rendezvous December 15, the day of the Gemini 6 launch. In a five-and-one-half hour period, Astronauts Schirra and Borman brought their spacecraft within six feet, close enough for the two crews to see each other through the spacecraft portholes.

The Gemini 6 then completed its 16-revolution flight, reentered the atmosphere and was recovered in the Atlantic Ocean. Meanwhile, the Gemini 7, which had been in space 12 days, completed its 14-day mission and was also recovered in the Atlantic.

Both astronaut crews emerged from their spacecraft in good condition. Astronauts Borman and Lovell, after two weeks, had traveled over five million miles in the space environment and given every indication that man can survive in space for extended periods.

The Gemini 7, at 17,500 miles an hour, completed 206 revolutions of the Earth in a 14-day period. This is more than the length of time the United States calculates is necessary to send men to the moon, land on its surface and return safely to Earth. The Gemini 6 completed 16 revolutions of the Earth during its two-day flight.

Astronauts Borman and Lovell were in space over 330 hours, establishing a new record for manned space flight.

Both spacecraft were launched by two-stage Titan II rockets. The high point (apogee) of the Gemini 7 was 203.6 statute miles and the low point (perigee) was 100 statute miles. Gemini 6 had an apogee of 161.5 statute miles and its perigee was 100 statute miles. Rendezvous was made at an altitude of 187.5 statute miles.

To achieve Earth orbit at these heights, a thrust of some 430,000 pounds from the first stage and 100,000 pounds of second stage thrust were required. The Gemini 7 weighed 8,076 and the Gemini 6 7,817 pounds.

The combined Gemini 7-6 missions were the fourth and fifth United States launches in the manned phase of the Gemini (from the twin bright stars, Castor and Pollux in the Gemini Zodiacal constellation) program. The first two Gemini launches were unmanned, and the following three were manned:

Gemini 3 – Astronauts Grissom and Young – March 23, 1965
Gemini 4 – Astronauts McDivitt and White – June 3-7, 1965
Gemini 5 – Astronauts Cooper and Conrad – August 21-29, 1965

In the program of 12 flights with five remaining, the Gemini 7-6 mission was of paramount importance in proving that long-duration missions are within the capability of both men and hardware.

Scientific experiments aboard the two spacecraft, most of which were on Gemini 7, included cardiovascular conditioning, inflight exerciser, phonocardiogram, human otolith test equipment, tri-axis magnetometer, celestial radiometry and space object radiometry equipment, space stadimeter and sextant, photometer, bioassay and calcium study equipment, Laser, photometric telescope sensors and photoelectric sensor.
It is too soon to determine conclusively how successful these experiments were, but preliminary data and

analyses pleased the doctors and scientists who are continuing examination of the great amount of data gathered. Results of the data studies will be made public in the United States and abroad.

THE LAUNCH VEHICLE – TITAN II

This was a two-stage rocket which stood 109 feet with the spacecraft aboard.

Its first stage was 63 feet high and the second stage 27 feet high. Both had a diameter of 10 feet.

Launch weight, including the spacecraft, was about 340,000 pounds.

The first stage had two rocket engines and the second stage had one.

The first stage engines produced a combined thrust at lift-off of 430,000 pounds, and the second stage produced 100,000 pounds of thrust at altitude (about 50 to 90 miles).

GEMINI SPACECRAFT

The Gemini spacecraft, looking much like its predecessor the Mercury (one-man) capsule, is conical in shape, 18 feet 5 inches long, 10 feet across at the base, and 39 inches across at the top. It is made up of the following components:

Reentry module – which includes the rendezvous and recovery sections, the reentry control system and the cabin.
Adapter Section – which contains the retrograde section and an equipment section.

(NOTE: The Gemini 7-6 mission was launched out of numerical sequence because the launch of the original October 25 Gemini 6 mission [rendezvous of the Gemini 6 spacecraft with an Agena target vehicle] was rescheduled when the Agena exploded on the launch pad. It was then decided to proceed with the Gemini 7 mission on schedule, and use it as a target vehicle for the rescheduled Gemini 6 launch December 15).
660215 #10003.5

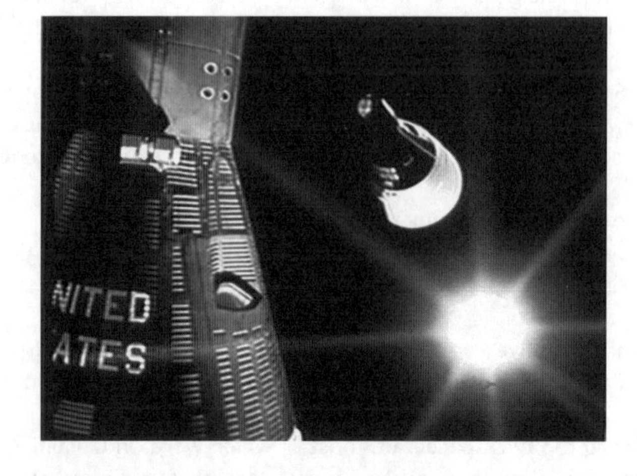

EDUCATIONAL BRIEF

National Aeronautics and Space Administration
Manned Spacecraft Center Houston, Texas

GEMINI SEVEN / SIX

The launching and rendezvous of Gemini VII and Gemini VI highlighted a very successful year for the Gemini program. The launch of two manned vehicles from the same launch pad in a short time and the setting of a number of new space flight records proved the United States manned space flight program to be highly operational.

GEMINI SEVEN

CAPTAIN JAMES A. LOVELL, PILOT
COLONEL FRANK BORMAN, COMMAND PILOT

At 10:48 c.s.t., December 4, 1965, astronauts Frank Borman, Command Pilot, and James A. Lovell, Jr., Pilot, entered the Gemini Seven spacecraft at complex 19, Cape Kennedy, Florida. The spacecraft doors were closed about 11:06 c.s.t., and Gemini Seven was launched with very little deviation from its programmed plan at 13:30 c.s.t.

The following comparison shows how close the Gemini Seven launch was to the planned program.

	Planned	Actual
Liftoff	13:30 c.s.t.	13:30:04 c.s.t.
Insertion velocity	25,804 ft./sec.	25,793 ft./sec.
Perigee	87 nautical miles	87.2 nautical miles
Apogee	183 nautical miles	177.1 nautical miles
Orbit plan inclination	28.87 degrees	28.89 degrees

About six minutes after liftoff and 450 miles downrange from Cape Kennedy, Gemini Seven attained an elliptical orbit. Gemini Seven separated from the second stage initially at about 2 feet per second, turned around immediately, thrusted back toward the tumbling second stage of the launch vehicle, and set up position about 50 to 60 feet away. Gemini Seven maintained this position with station keeping maneuvers for about 20 minutes while taking pictures and making measurements.

Astronauts Borman and Lovell maintained a regular 24-hour cycle or day using Central Standard Time. After the first night, both astronauts slept at the same time. This was done because crews on previous flights indicated that sleep was interrupted by disturbances produced by the astronaut who was awake during alternate sleep cycles.

The Gemini Seven astronauts wore new lightweight spacesuits which Lovell had helped to develop. On the second day, Lovell removed his spacesuit and was clad only in an undergarment similar to long underwear.

The Gemini 7 crew conducted 20 in-flight experiments, 14 of which were continuations of experiments conducted on previous Gemini flights. The new experiments were (1) Bioassays of Body Fluids; (2) Calcium Balance Study; (3) In-flight Sleep Analysis; (4) Optical Communication (laser); (5) Landmark Contrast Measurements; (6) Star Occultation Navigation.

At 3 hours 20 minutes into the flight, Gemini Seven successfully sighted on the star Spica for an orbital maneuver. This was the first time that visual sighting on a star had been used for a spacecraft maneuver. Gemini Seven circularized its orbit to about 161 miles in three steps in preparation for rendezvous with Gemini 6.

A good example of the level of visual perception attained in the flight is the fact that the Gemini 7 crew observed and tracked a Minuteman missile over the Pacific Ocean with an estimated rate of closure between the two vehicles of 29,000 miles per hour. The crew also made radiometric measurements of the Minuteman missile as it reentered in the area over Kwajalein.

GEMINI SIX

MAJOR THOMAS STAFFORD, PILOT
CAPTAIN WALTER SCHIRRA, COMMAND PILOT

Gemini Six, with Astronauts Walter M. Schirra, Command Pilot, and Thomas P. Stafford, Pilot, was launched at 7:37 c.s.t. December 15, 1965, from complex 19 at Cape Kennedy, Florida. The Gemini Six launch was made as Gemini 7 passed over Cape Kennedy. The launch time of Gemini 6 was of utmost importance in order to achieve rendezvous with Gemini 7. Gemini 6 experienced a launch as close to plan as had Gemini 7.

The following comparison shows how close performance came to the planned program.

	Planned	Actual
Ignition	7:37:83 c.s.t.	7:37:23 c.s.t.
Liftoff	7:37:26 c.s.t.	7:37:26 c.s.t.
Insertion velocity	25,730 ft./sec.	25,718 ft./sec.
Perigee	87.1 nautical miles	87.2 nautical miles
Apogee	146.2 nautical miles	140.0 nautical miles
Inertial Period	88 min. 46 sec.	88 min. 55 sec.
Orbit plan inclination	28.87 degrees	28.97 degrees

After separation from the second stage of the launch vehicle, Gemini 6 made an 11½-second posigrade burn to place it in orbit. Gemini 7 led Gemini 6 by 1,200 nautical miles at the time of insertion, or about 5 minutes in time. Gemini 6, in a slightly lower orbit, had a shorter period and a slightly higher velocity, and closed this distance gradually during the orbits preceding adjustment and final rendezvous. Rendezvous was predicted at 5 hours 48 minutes and 51 seconds into the flight of Gemini 6, in an area northwest of Guam over the Marianna Islands.

Gemini 6 made several height adjustments to place it in an orbit of about 146 miles altitude and one plane adjustment to place it the same plane as Gemini 7. A terminal phase burn raised Gemini 6 to the same orbit level as Gemini 7, and rendezvous was achieved as scheduled. Astronaut Stafford's report of one hundred twenty feet apart and steady was the signal that rendezvous had been achieved. An interesting coincidence during rendezvous was the report of the Gemini 6 crew that the Gemini stars, Castor and Pollux, could be seen just to the right of Gemini 7.

Gemini 6 and 7 then orbited the Earth within close range of each other, closing at one time to approximately one foot apart. Gemini 6 also performed an in-plane and out-of-plane fly around Gemini 7.

GEMINI VII AS VIEWED FROM GEMINI VI

During the 8th revolution of Gemini 6 and the 169th revolution of Gemini 7, Gemini 6 changed its orbit to 163.2 miles apogee and 154.1 miles perigee. The Gemini 7 orbit remained at 163.7 miles apogee and 159.0 miles perigee, causing Gemini 6 to move slightly ahead of Gemini 7. These orbits were maintained during the sleep period and the two separate spacecraft varied from 22 to 42 miles apart. Gemini 6 and 7 maintained position within 100 kilometers (about 60 miles) of each other for a record of 20 hours and 22 minutes.

At 8:53:24 c.s.t., Gemini 6 fired its retro-rockets in the area about 700 miles northwest of Canton Island. At 9:29:09 c.s.t. December 16, 1965, Gemini 6 splashed down in the Atlantic Ocean about 12 miles southwest of the aircraft carrier Wasp. The Gemini 6 crew elected to remain inside their spacecraft until it was lifted aboard the carrier.

The Gemini 7 crew meanwhile experienced difficulty with their number 3 and 4 thrusters (the right yaw thrusters) and were unable to track Gemini 6 during retrofire and reentry.

During its 206th revolution, at 7:28:01 c.s.t. December 18, 1965, Gemini 7 fired its retrorockets over the equator, about 3,000 miles east of the Philippines. Splashdown was at 7:59:10 c.s.t. about 8 miles northwest of the aircraft carrier Wasp.

MOON AS VIEWED FROM GEMINI VII

The flights of Gemini 7 and 6 demonstrated ability to rendezvous and established a number of new space records as well. The following chart is a summary of space flight records to the date shown.

Space flight records to January 1, 1966

	Spacecraft Time	Man-hours in space
Soviet (Total)	432 hours 40 minutes	507 hours 16 minutes

	Spacecraft Time	Man-hours in space
United States		
Before Gemini 7/6	346 hours 39 minutes	641 hours 24 minutes
Gemini 6	26 hours 01 minutes	52 hours 03 minutes
Gemini 7	330 hours 35 minutes	661 hours 11 minutes
Total	703 hours 15 minutes	1354 hours 38 minutes

NATIONAL AERONAUTICS AND SPACE ADMINISTRATION
WASHINGTON, D.C. 20546

RELEASE NO: 65-347

FOR RELEASE: IMMEDIATE
November 9, 1965

NASA SCHEDULES GEMINI 7 LAUNCH NO EARLIER THAN DECEMBER 4; GEMINI 6 DECEMBER 13

The National Aeronautics and Space Administration today announced the launch of Gemini 7 – first of two launches in a combination long-duration mission and rendezvous of two manned Gemini spacecraft – is scheduled no earlier than December 4.

If preparation of launch facilities and checkout of launch vehicle and spacecraft proceed as presently planned, the launch of Gemini 6 will follow nine days later on December 13.

James A. Lovell and Frank Borman

Astronauts Frank Borman and James A. Lovell, Jr., are command pilot and pilot respectively for the Gemini 7 mission. It will be the first space flight for each. Astronauts Edward H. White, II, and Michael Collins are the backup crewman.

Walter M. Schirra is the command pilot of Gemini 6 and Thomas P. Stafford is the pilot. It will be the second space flight for Schirra. Astronauts Virgil I. Grissom and John W. Young are the backup pilots.

The Gemini 7 mission is scheduled for up to 14 days. The purpose of this flight is to determine the effects of long duration flight on man. Twenty scientific, medical and technological experiments are scheduled to be carried out on the Gemini 7 mission.

The mission plans for Gemini 6 are nearly identical to those of the original rendezvous flight which was postponed October 25 when the Agena Target Vehicle failed to achieve orbit. (An intensive study is now underway to determine the cause of the Agena failure.)

Gemini 6 will rendezvous with the Gemini 7 target spacecraft during the fourth revolution and station-keep with the Gemini 7 spacecraft for a period of time. Maximum duration of the Gemini 6 mission is two days. While the Gemini 6 spacecraft will approach within close proximity of the Gemini 7 spacecraft, it will not dock with it.

Michael Collins and Edwars H. White, II

No major changes have been made in the Gemini 7 mission. The Gemini 6 mission will have no major impact on the accomplishment of Gemini 7 objectives. The Gemini 7 flight trajectory has been modified to provide support as a target for the Gemini 6 rendezvous mission.

The purpose of proceeding with the attempt to launch Gemini 6 is to demonstrate as early as possible a rendezvous of two vehicles in space.

NATIONAL AERONAUTICS AND SPACE ADMINISTRATION MANNED SPACECRAFT CENTER
Houston, Texas

MSC 65-109
November 24 , 1965

HOUSTON, TEXAS – The MSC remote site flight controller teams for the Gemini 7/6 mission began deploying this week to the seven locations around the world where they will exercise detailed real time mission control during the upcoming flights of the two Gemini spacecraft.

In addition to the six remote sites: Canary Islands (CYI); Canarvon, W. Australia (CRO); Kauai, Hawaii (HAW); Guaymas, Mexico (GYM); tracking ship Rose Knot (RK); and tracking ship Coastal Sentry (CS); a crew from Houston will also man the Corpus Christi Texas, (TEX) site.

The teams, each composed of from four to seven men, are scheduled to be on station by noon Thanksgiving Day to begin preparing the sites and crews for the network simulations prior to the Gemini 7/6 mission.

A pre-mission preparation phase will be the first order of business at each remote site. The first day, briefings will be given the local maintenance and operations (M&O) people by the flight controllers.

This will consist of briefings on the general mission flight plan, and a discussion of procedural changes that have been instituted since the previous mission. The senior flight controller who is the Command Communicator at each site will give the briefing.

The Gemini systems engineer will brief the M&O group on the Gemini spacecraft and their unique aspects for the current mission and any specialized "backroom" monitoring procedures deemed appropriate for a given spacecraft pass or the entire mission.

A briefing by the aeromedical monitor will be held to provide the M&O staff with the medical aspects of the mission and biomedical or other research experiments to be performed.

Another member of the team at the remote sites is the astronaut simulator. His duties prior to the actual mission is to man the astronaut simulator console at the remote site and play the part of the astronauts in the spacecraft. He also controls the pre-mission remote site simulations. During the mission he will perform duties as backup command communicator or spacecraft systems engineer at the site.

On the second day a network test simulation will be held at the remote site to confidence test the equipment. This will be followed by network tests to integrate the remote site into the network operation. Local confidence tests are run to integrate the remote site flight controllers and the M&O people and at the same time develop confidence in the remote site systems. A state of readiness is maintained at each site until liftoff on mission day.

Prior to deployment, the remote site crews study the standard operating procedures for the remote sites such as the use of the command system and telemetry, the air-to-ground communications, network tests and other necessary procedures. Each man gets at least three hours in the Gemini procedures trainer for cockpit familiarization. They also attend systems and procedural briefings.

In the days before deploying to the remote sites the crews take part in simulated network simulations in the Mission Control Center. From a back room in the MCC, the remote site flight controllers conduct simulated missions as though they were on station at the various sites.

The command communicator at the remote site, in addition to supervising the preparation of the site for the mission, is the delegated representative of the Mission Control Center Flight Director and serves as the operations manager of the site and its supporting crew throughout the mission phase. He is responsible for the air to ground communications with the spacecraft and for operation of the ground to spacecraft command system. Any decision from a site that affects the mission is made by the command communicator.

Two Gemini systems engineers will be at each site and their area of responsibility is to monitor, analyze and report any spacecraft systems anomalies noted on the telemetry displays. They in turn will make recommendations for corrective action to the command communicator and/or the MCC-H.

Each site will have two surgeons (aeromedical monitors) whose primary responsibility is to monitor the physical and physiological well being of the astronauts via telemetry. The information is reported by the doctors to the command communicator who relays the information to the MCC-H.

During the course of the mission the command communicator normally relays all information to the MCC-H and the spacecraft crew. The exception being when a discussion with the flight controller's counterpart at the MCC-H is necessary to resolve a point.

The normal work day for the remote site flight controllers is 14 to 15 hours. The hours worked during each 24-hour period are determined by lift-off time of the spacecraft from Cape Kennedy, which in turn sets the ground track passes over each station. An average of seven passes are over each remote site daily.

During periods when the spacecraft is not over the site the operation goes on standby. Then about two and one-half hours before acquisition time on the first pass of a series of passes by the spacecraft, all equipment at the site is confidence tested and all mission teletype messages are reviewed in preparation for support of the mission.

Flight controllers at the remote sites stay in commercial facilities as near the sites as possible. Limited food facilities are available at all remote stations and the crews usually get in a little cooking experience also.

Team member assignments vary from mission to mission and a flight controller may be assigned to any one of the seven sites for a Gemini flight. With the present schedule of Gemini flights, many of the flight controllers are on travel nearly half of the time. Most of the men are married and would like to spend more time with their families but the fact that they do enjoy their work helps in some small way to help compensate for the time away from home.

After the mission is completed and the crews return to Houston, they conduct a complete evaluation of the operation of equipment and procedures used at the remote site and make necessary recommendations for hardware changes that are deemed necessary to improve operations. They also evaluate all documentation used and try to make improvements for the next mission.

Flight controllers for the Gemini 7/6 mission and their stations are as follows:

Canary Islands (CYI) site crew is James R. Fucci, command communicator (CC); Floyd E. Claunch and Robert D. Legler, Gemini systems (GS); Luis J. Espinoza, astronaut simulator (AS); and Lt. Col. John W. Ord, USAF and Cant. Charles Wilson, USAF, aeromedical monitors (AM).

Carnarvon, W. Australia (CRO) site members are Keith K. Kundel (CC); James F. Moser and George M. Bliss (GS); Edward L. Dunbar (AS); and Capt. Edward L. Beckman, USN, Sqdn. Ldr. Mike Mury/Alston and Wing Cdr. L. N. Walch (AM).

Kauai, Hawaii (HAW) is staffed by Edward I. Fendell and Capt. William F. Buchholz, USAF (CC); Hershel R. Perkins and Joseph Fuller, Jr. (GS); John W. Collins (AS); and Cdr. Eustache Prestcott Jr., USN, and Maj. James R. Wamsley, USAF (AM).

Guaymas, Mexico (GYM) site staff for the Gemini 7/6 mission is Gary B. Scott (CC); George W. Conway and Albert W. Barker (GS); Harold V. Berlin (AS); and Maj. Richard M. Chubb, USAF and Maj. William P. Nelson, USAF (AM).

The tracking ship Coastal Sentry (CS) to be on station northeast of Luzon, Philippines, will be manned by Charles R. Lewis and Harold M. Draughon (CC); Gene F. Muse and Harry Smith (GS); Willard D. Robinson (AS); and Maj. Joseph A. Ionno and Cdr. Robert W. Maher, USN (AM).

Flight controllers onboard the tracking ship Rose Knot (RK) located in the Atlantic Ocean northeast of Victoria, Brazil, will be Willima D. Garvin (CC); Charles A. Link and John E. Walsh (GS); James R. Bates (AS); and Maj. O'Neill Barrett, USAF and Cdr. Michael C. Carver, USN (AM).

Corpus Christi, Texas (TEX) site will be manned by Arda J. Roy, Jr. and Maj. William G. Bastedo, USAF (CC); Dale L. Klingbeil (GS); and Capt. Lawrence J. Enders, USAF (AM).

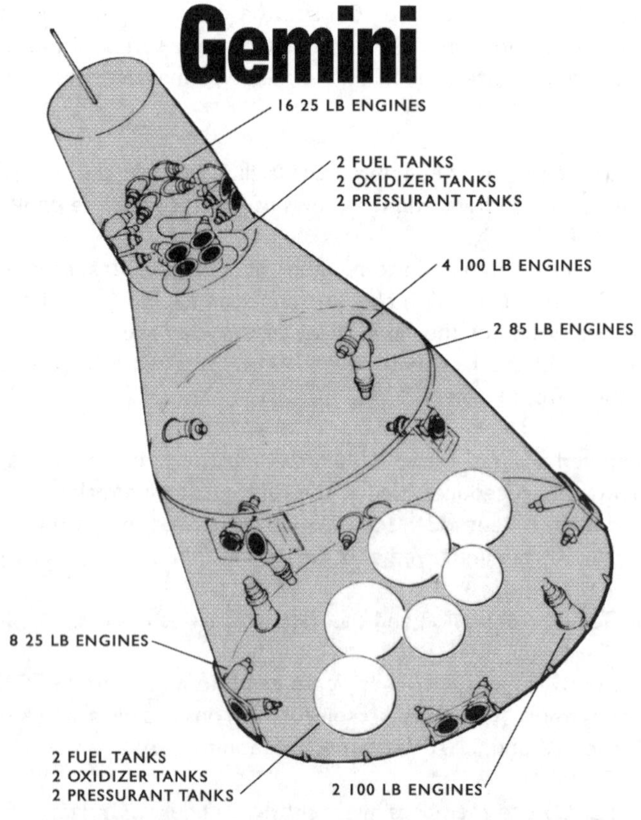

GEMINI CAPSULE MANEUVRING ENGINE AND TANKS LOCATIONS

**NATIONAL AERONAUTICS
AND SPACE ADMINISTRATION
WASHINGTON, D.C. 20546**

FOR RELEASE: MONDAY P.M.
November 29, 1965

RELEASE NO: 65-362

P R E S S K I T

PROJECT: GEMINI 7/6

GEMINI 7 / 6 FLIGHT TO ATTEMPT RENDEZVOUS, LONG DURATION MISSIONS

Within the next three weeks, the National Aeronautics and Space Administration is scheduled to carry out two manned space missions – a long-duration flight of up to 14 days and rendezvous of two Gemini spacecraft.

Gemini 7, a long-duration flight, is scheduled to be launched no earlier than December 4, 1965.

Gemini 6, which will rendezvous with Gemini 7, is to be launched nine days later, December 13, 1965.

Success in the two flights will represent:

1. The longest U.S. manned flight to date (Gemini 5 Astronauts L. Gordon Cooper and Charles Conrad were in flight 190 hours and 56 minutes, nearly eight days).
2. The first space rendezvous of two manned maneuverable spacecraft.
3. A minimum turn-around time for launch of two missions from the same pad.

Despite the rendezvous objective of Gemini 6, the two missions will be carried out independently. That is, Gemini 7 will be launched and carried out as originally planned. No major changes have been made in the Gemini 7 flight plan.

The Gemini 6 mission will be carried out according to a flight plan which is nearly identical to the one prepared for the October 25 launch which was postponed when the Agena Target Vehicle failed to achieve orbit. The only major change is that the Gemini 6 spacecraft was to have docked with the Agena.

In the forthcoming flight, the two Gemini spacecraft will not be physically connected.

It was decided that the planned schedule of Gemini 7 and the availability of the Gemini 6 launch vehicle and spacecraft (already checked out on Pad 19) presented an opportunity to carry out a rendezvous of two manned vehicles.

The first several days of the Gemini 7 mission will be devoted to carrying out experiments. At about five days the crew will maneuver the spacecraft into a target orbit for Gemini 6. During the rendezvous attempt by Gemini 6, the Gemini 7 crew will maintain their orbit and spacecraft attitude, performing only those maneuvers required to make themselves a better target.

Following a successful liftoff, Gemini 6 immediately will begin maneuvers to achieve rendezvous which is planned for its fourth orbit. Following rendezvous, Gemini 6 will station keep (fly formation) on Gemini 7 for about two revolutions. Subsequently, Gemini 7 will fly formation on Gemini 6 for one spacecraft day, about 40 minutes.

Gemini 6 will reenter the Earth's atmosphere and land in the West Atlantic Ocean after about 46 hours and 45 minutes, at approximately 8:20 a.m. EST.

Gemini 7 duration will be about 329 hours and 30 minutes, landing in the same area at approximately 8 a.m. EST, two days later.

(BACKGROUND INFORMATION FOLLOWS)

LAUNCH VEHICLE COUNTDOWN

Gemini 7	
F-3 days	Start pre-count
F-1 day	Start mid-count
T-12 hours	GLV propellant loading
T-390 minutes	Complete propellant loading
T-300 minutes	Begin terminal countdown
T-120 minutes	Flight Crew to Complex 19
T-100 minutes	Crew enters spacecraft
T-75 minutes	Close spacecraft hatches
T-50 minutes	White Room evacuation
T-35 minutes	Begin erector lowering
T-15 minutes	Spacecraft OAMS static firing
T-04 seconds	GLV ignition
T-0 seconds	Liftoff
T+2 minutes, 36 seconds	Booster engine cutoff (BECO)
T+5:41	Second stage engine cutoff (SECO)
T+6:11	Spacecraft-launch vehicle separation

NOMINAL MISSION PLAN

Gemini-7

Gemini-7 is scheduled to be launched from Complex 19, Cape Kennedy at about 2:30pm EST, December 4. It will be launched into an elliptical orbit with an apogee of 210 miles and a perigee of 100 miles. The orbit will be inclined 28.87 degrees to the equator.

The spacecraft is to separate from the booster 30 seconds after sustainer engine cutoff.

Immediately following spacecraft separation, the spacecraft will turn around to blunt end forward, and begin station keeping on the booster second stage. Station keeping will continue for about 25 minutes ground elapsed time (GET) from liftoff. This time extends into about five minutes before the first darkness period.

Celestial radiometry experiments will be conducted during the remainder of the first darkness period.

At three hours, 50 minutes after lift-off as the spacecraft is at its third apogee, thrusters will be fired in a posigrade maneuver to raise the perigee to 124 miles. This maneuver establishes a spacecraft orbital lifetime of 15 days.

The next several days of the flight will be devoted to conducting assigned experiments.

At about five days in flight the crew will circularize the spacecraft orbit to provide the proper target orbit for Gemini 6. The exact maneuvers required will depend on the decay rate of the Gemini 7 orbit and the expected liftoff time of Gemini 6, now planned for eight days, 19 hours and four minutes following Gemini 7 liftoff.

Under present plans, circularization maneuvers will be performed 120 hours after lift-off of the Gemini 7 spacecraft. The crew will give the spacecraft a posigrade thrust at apogee. This will result in a change of velocity of 100 feet per second and circularization at 185 miles.

The rendezvous portion of the Gemini 6 and Gemini 7 spacecraft is described in the Gemini 6 Nominal Mission Plan.

The remaining Gemini 7 experiments will be conducted following completion of rendezvous activities. The Gemini 7 crew will initiate retrofire near the end of the 206th revolution. Landing will be in the West Atlantic recovery area at the beginning of the 207th revolution.

GEMINI 7 EXPERIMENTS

Twenty experiments are scheduled for Gemini 7. Fourteen are continuing experiments and have been carried aboard previous Gemini flights. In repeating these over a number of manned space flight missions, experimenters hope to gain data covering a number of subjects under varying flight conditions.

Results of the experiments carried aboard Gemini missions 3 and 4 were presented at a symposium in Washington early this fall. Similar symposia will be held periodically during the manned space flight program.

Experiments flown on Earlier Missions

1. Cardiovascular Conditioning (5)*

   This experiment will determine the effectiveness of cyclic inflations of pneumatic cuffs on the thighs as a preventive measure of cardiovascular deconditioning (heart and blood distribution system) induced by prolonged weightlessness. The cuffs are built into the spacesuit around the thighs and inflated periodically to 80mm of mercury pressure, increasing blood pressure below the cuffs. The automatic pressurization cycle lasts two minutes out of every six and uses oxygen from the environmental control system.

* indicates previous Gemini mission

2. In-Flight Exerciser (4 and 5)

   The objective of this experiment is to assess the astronauts' capacity to perform physical work and their capability for sustained performance. The rapidity with which the heart rate returns to normal after cessation of exercise is an indication of an individual's physical fitness. A workload will be provided by specific periods of exercise at the rate of one pull per second for 30 seconds on an exercise device that requires a known amount of effort.

   The exercise device consists of a pair of bungee cords attached to a nylon handle at one end and a nylon foot strap at the other end.

   The in-flight data obtained will be compared with the control data to determine the capacity for work in space.

3. In-flight Phonocardiogram (4 and 5)

In this experiment the fatigue state of an astronaut's heart muscle will be determined by measuring the time interval between the activation of a muscle and the onset of its contraction. A microphone will be applied to an astronaut's chest well at the cardiac apex. Heart sounds detected during the flight will be recorded on an onboard biomedical recorder. The sound trace will be compared to the waveform obtained from a simultaneous inflight electrocardiogram to determine the time interval between electrical activation of the heart muscle and the onset of ventricular systole.

4. Bone Demineralization (4 and 5)

The purpose of this experiment is to establish the occurrence and degree of bone demineralization influenced by the relative immobilization associated with the cockpit of the Gemini spacecraft and weightlessness.

Special X-rays will be taken of an astronaut's heel bone and the terminal bone of the fifth digit of the right hand. Three pre-flight and three post-flight exposures will be taken of these two bones and compared to determine if any bone demineralization has occurred due to the space flight.

The equipment to be used in this experiment will be closely calibrated clinical X-ray machines, standard 11-inch by 14-inch X-ray films and calibrated wedge densitometers.

5. Human Otolith Function (5)

A visual tester will be used to determine the Astronaut's orientation capability during flight. The experiment will measure changes in otolith (gravity gradient sensors in the inner ear) functions.

The tester is a pair of special light proof goggles, one eye piece of which contains a light source in the form of a movable white line. The astronaut positions the white line with a calibrated knurled screw to what he judges to be the right pitch axis of the spacecraft. The second astronaut then reads and records the numbers.

6. Proton-Electron Spectrometer (4)

In order to determine the degree of hazard, if any, to which the crew will be subjected on space flight, it is necessary to project what radiation environment any given mission will encounter. Specifically, this experiment will make measurements outside of the spacecraft in a region where the inner Van Allen radiation belt dips close to the Earth's surface due to the irregular strength of the Earth's magnetic field. This region is usually referred to as the South Atlantic Geomagnetic Anomaly.

This measurement will be accomplished by means of a scintillating-crystal, charged-particle analyzer mounted on the adapter assembly of the spacecraft. Data from this experiment will be used to correlate radiation measurements made inside the spacecraft and to predict radiation levels on future space missions.

7. Tri-Axis Magnetometer (4)

The purpose of this experiment is to monitor the direction and amplitude of the Earth's magnetic field with respect to an orbiting spacecraft. The astronauts will operate an adapter-mounted tri-axis flux gate magnetometer as they pass through the South Atlantic Geomagnetic Anomaly. The magnitude of the three directions of the Earth's magnetic field will be measured with respect to the spacecraft. The measurement will be performed in conjunction with the Proton-Electron Spectrometer experiment to determine the field, line direction and pitch angle of the impacting particles.

8. Celestial Radiometry / Space Object Radiometry (5)

The results of these experiments will provide information on radiation intensity of celestial bodies and various objects in space. Instrumentation includes a three-channel spectro-radiometer, a dual-channel Michelson Interferometer-Spectrometer, and a cryogenically cooled spectrometer. The equipment can measure radiant intensity from the ultra-violet through the infrared region. The sensing units are housed in the Gemini adapter section and are directed toward the objective by orienting the spacecraft.

The objectives for these experiments are to determine the onset of sensitivity values for Earth objects and sky background radiation and radiation signatures of various objects in space and on the ground. Observations will include exhaust plumes of rocket vehicles launched from the Eastern or Western Test Ranges, rocket sled exhausts at Holloman Air Force Base, volcanoes and forest fires as well as contrasting background areas such as deserts and warm ocean currents.

9. Simple Navigation (4)

The capability of man to navigate in space and to provide a reliable navigation system independent of ground support will be tested in this experiment. Two special instruments have been developed for use on Gemini spacecraft to allow detailed manual-visual examination of the space phenomena thought to be best for space navigation purposes. These are a space stadimeter and a sextant. This flight will only carry the space sextant which the astronaut will use to make star-horizon angular measurements for orbital orientation determinations. The results will be compared with actual measurements to determine the accuracy of the procedures.

10. Synoptic Terrain Photography (4 and 5)

The purpose is to obtain photos of selected parts of Earth's surface for use in research in geology, geophysics, geography, oceanography. This experiment has been flown on every flight since MA-8.

Experiment – 70mm Hasselblad camera with 80mm Zeiss F2.8 lens; two packs of color film with 65 exposures each. Approximately ninety pictures will be taken over areas of the world. Primary areas are the shallow waters around the Bahamas, The Red Sea, and the west central portion of Mexico.

11. Synoptic Weather Photography (3, 4 and 5)

The Synoptic Weather Photography experiment is designed to make use of man's ability to photograph cloud systems selectively – in color and in greater detail than can be obtained from the current TIROS meteorological satellite.

A primary purpose of the experiment is to augment information from meteorological satellites which are contributing substantially to knowledge of the Earth's weather systems. In many areas they provide information where few or no other observations exist.

Experiment – 70mm Hasselblad camera with 80mm Zeiss F2.8 lens; two magazines of color film with 65 exposures each. Areas of Interest – Squall line clouds, thunderstorm activity not associated with squall lines, frontal clouds and views of fronts, jet stream cirrus clouds, typical morning stratus of Gulf states, coastal cloudiness, tropical and extratropical cyclones, intertropical convergence zone, cellular pattern in subtropical phenomenon, wave clouds induced by islands and mountain ranges, broad banking of clouds in the trade winds or other regions.

12. Visual Acuity / Astronaut Visibility (5)

The visual ability of the astronauts in the detection and recognition of objects on the Earth's surface will be tested in these experiments. The spacecraft will be equipped with a vision tester and a photometer.

The astronauts will use the vision tester to evaluate visual sightings from space relative to Earthbound baseline values.

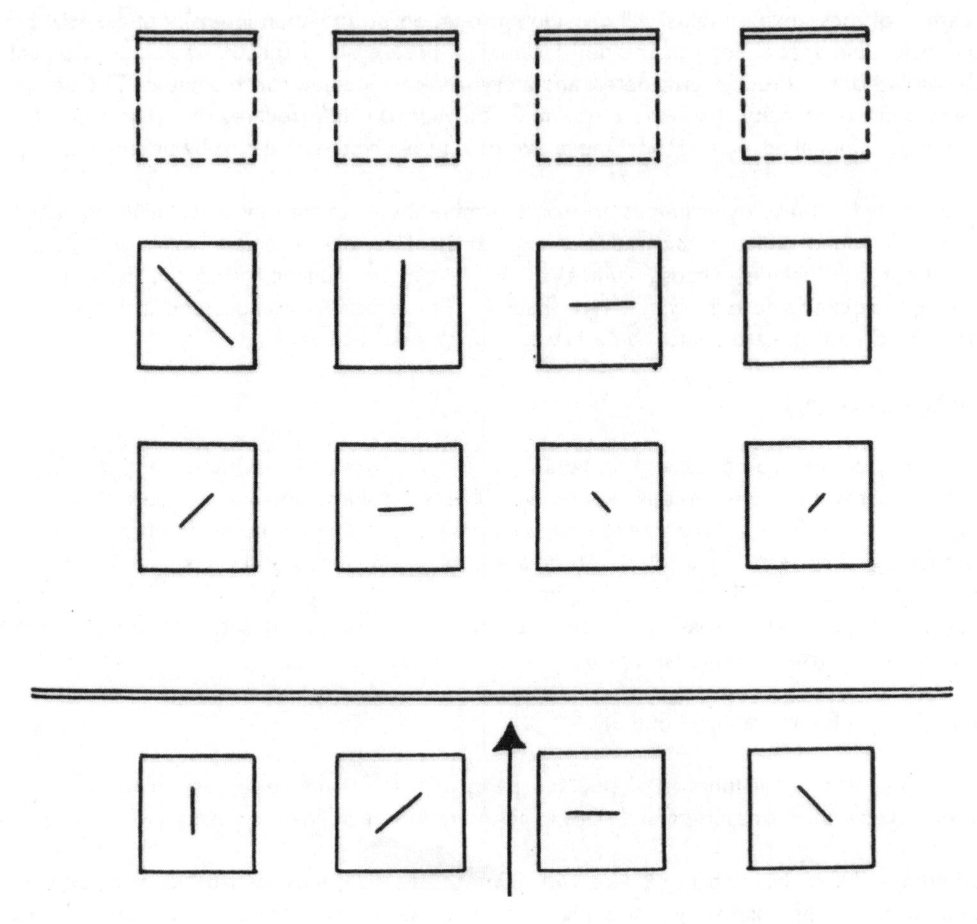

The photometer will measure light attenuation of the spacecraft window due to scattering. While the spacecraft is oriented the astronauts will view a pattern of panels laid out near Laredo, Texas and record their findings. Viewings will be correlated with laboratory experiments and vision will be checked pre- and post-flight.

During passage of the spacecraft over the sites, the command astronaut shall be responsible for maintaining the proper spacecraft attitude while the second astronaut observes the target area and makes verbal comments to the principal investigator at the site.

For five minutes in each 24 hour period, each astronaut will use the on-board vision tester to test his own visual acuity on an opportunity basis.

Experiments To Be Flown For The First Time

1. Bioassays Body Fluids

In this experiment the astronauts' reaction to stress during space flight will be studied by means of analyzing body fluids. Pre-flight and post-flight blood samples will be taken. In-flight urine will be measured at each voiding and a portion of this stored in special bags.

From analysis of these fluids experimenters hope to measure body hormones, electrolytes, proteins, amino acids and enzymes which may be produced as a result of stress.

2. Calcium Balance Study

The rate and amount of calcium change to the body during the conditions of orbital flight will be evaluated in this experiment by means of controlled calcium intake and output measurements. In addition to calcium, other electrolytes of interest such as nitrogen, phosphorous, sodium chloride and magnesium will be monitored.

The two astronauts will be maintained on a prescribed calcium diet for two weeks prior to flight, during and after flight. Careful recording of input-output will be accomplished and total fecal and urine specimens will be preserved for analysis. Sweat will also be measured by careful cleansing of the crew in distilled water following recovery. Undergarments will be similarly cleaned and the water analyzed.

3. In-flight Sleep Analysis

The objectives of this experiment are to assess the astronauts' state of alertness, levels of consciousness, and depth of sleep during flight. An electroencephalograph (EEG) on the astronauts will be taken during weightless flight to establish the possible use of the EEG as a monitoring tool to help determine the state of alertness and depth of sleep. The electrical activity of the cerebral cortex will be monitored by two pairs of scalp electrodes and recorded on the biomedical recorder.

4. Optical Communication (Laser)

Laser is an acronym for Light Amplification by Stimulated Emission of Radiation. Stimulated emission is produced by greatly exciting the atom. When excited, the atom will emit small quantities of light in phase or unison. Thus the light is "coherent," that is, it is directed in a constant steady beam in one precise direction.

This experiment is an attempt to demonstrate a new technique for communication between an orbiting spacecraft and a ground station. In doing this, a demonstration of optical frequencies for communications will be achieved and certain atmospheric data will be recorded and the value of an astronaut as a "pointing control" will be established.

The primary atmospheric data to be obtained are background radiance and attenuation. This data and the experience obtained from this experiment will be useful in designing future systems.

The experiment equipment consists of a flight transmitter and a ground-based receiver transmitter system.

The flight transmitter resembles and is about the same size as a home movie camera. It weighs about six pounds and is completely self-contained. It is made up of four injection lasers, a 10-volt power supply (eight rechargeable nickel-cadmium batteries), a D.C. (direct current) to D.C. converter, a telescopic sight and a microphone.

Four gallium arsenide injection lasers are the heart of the transmitter. They deliver a total of 16 watts of light power at a wavelength of 9,000 angstroms. The beams produced by the lasers form four lines of light arranged one above the other making a square pattern at distances of several feet to infinity.

Injection lasers were chosen for their compactness, light weight and efficiency in converting electric energy into light energy.

The ground-based receiver resembles a short, blunt telescope. It is 30 inches in diameter and consists of a collector and focusing unit with a photomultiplier (optical detector) located at the focal plane. An argon gas laser beacon is mounted atop the receiver barrel.

The argon gas laser has an output of three watts into a three milli-radian beam spread or about 0.17 degrees. At 300 miles the beam will be approximately 0.9 miles in diameter.

Receiver systems have been installed at White Sands Missile Range, Ascension Island and Kauai, Hawaii. The receivers are slaved to FPS radars and always point toward the spacecraft when it is within range of the radar.

In operation the command pilot will maintain proper spacecraft orientation while the co-pilot aims the laser transmitter by sighting through the telescope at the ground-based argon laser. The argon laser beam will be visible to the naked eye.

When the spacecraft laser beacon is acquired by the ground receiver the ground-based argon laser will be flashed to indicate that contact has been established. Both beacons will then be aligned and voice communications can begin. The co-pilot will switch to the voice channel and say, "1, 2, 3, 4, 5, testing 5, 4, 3, 2, 1."

Voice communications will be one way only – from spacecraft to ground.

During the experiment the astronauts will wear safety goggles for protection against eye damage which might be caused by stray or reflected light from the onboard laser. The glasses have shields for stopping radiation which enter the eye from the side and lenses that filter out the infrared energy emitted by the laser.

5. Landmark Contrast Measurements

The purpose of this experiment is to measure the visual contrast of land-sea boundaries and other types of terrain to be used as a service of navigation data for the onboard Apollo Guidance and Navigation system. Landmark contrast measurements made from outside the atmosphere will provide data of a high confidence level to effectively duplicate navigation sightings for Apollo.

Landmark measurements will be made of such areas as the Florida Coast, South American Chilean Coast, African-Atlantic Coast and Australian Coast. A photometric telescope sensor and equipment used for the Star Occultation Navigation experiment will be used.

6. Star Occultation Navigation

The feasibility and operational value of star occulting measurements in the development of a simple, accurate and self-contained orbital navigational capability will be investigated in this experiment.

The astronauts will determine the orbit of the Gemini spacecraft by measuring the time stars dip behind an established horizon.

As much of the existing Gemini onboard equipment as is possible will be used for the recording of photometric sensor output signal intensity and time. Nevertheless, certain special equipment will be necessary for the performance of the navigational studies. Included in the equipment is a photoelectric sensor.

The photoelectric sensor consists of a telescope, eyepiece, reticule, partially silvered mirror, iris, chopper, optical filters, photomultiplier, pre-amplifier and associated electronics. The instrument is hand-held to the astronaut's eye for viewing out the spacecraft's window.

As the astronaut views the horizon, he looks for bright stars about to be occulted. He then points the telescope at one and centers the star within a reticule circle. A portion of the radiation is then diverted to a photomultiplier. With a hand-held switch, the astronaut initiates a calibration mode in which the intensity of the star is measured automatically. He then tracks the star within the reticule as the star passes into the atmosphere and behind the edge of the Earth. The tracking period for each star is approximately 100 seconds. During this tracking period the astronaut will manually indicate the passage of the star through the air glow, the 50% intensity level and complete occultation simply by momentarily depressing the calibration switch.

The astronaut plays an essential role in the procedure, First, he solves the star acquisition problem by locating the next star to be transited. Second, he uses his head to point the telescope thus eliminating a two-gimbal automatic tracking system which would of necessity be used if he were not onboard. Third, he records star occultation times manually for comparison with the automatic calibration mode. Finally he notes peculiarities in the data as it is collected. In performing this latter function, the man is used to greatest advantage to advance the state of the navigational art as rapidly as possible.

CAMERA EQUIPMENT FOR GEMINI 7 AND 6 MISSIONS

16mm Maurer Movie Camera

I. Camera
   A. Equipment
      1. two cameras
      2. 75mm lens (one camera)
      3. 75mm, 25mm, 18mm lens set (second camera)
   B. Characteristics
      1. Six frames/second
      2. f-11 aperture
      3. 1/200 second shutter speed
      4. 40 lines/mm resolution
II. Film
   Kodak S.O. 217 color film
III. Purpose
   Weather and Terrain Photography
   General Purpose

70mm Hasselblad Camera

I. Camera
   A. Equipment
      1. Camera
      2. 80mm lens
      3. 250mm lens
      4. Photo event indicator
      5. Ring Sight
      6. UV filter
      7. Film backs
   B. Characteristics
      1. 80mm focal length
      2. f2.8 to f22.0 aperture
      3. Time exposures and speeds up to 1/500 second
      4. Resolution: approximately 125 lines/mm
      5. Approximately 1.5X magnification
II. Film
   Kodak S.O. 217, MS Ektachrome
   ASA-64 color emulsion on 2.5 mil Estar Polyester base
III. Purpose
   Weather and Terrain
   General Purpose

GEMINI 7 – FOURTEEN DAY MENU CYCLE

MENU I – DAYS 1, 5, 9 & 13

Meal A	(Days 5, 9, 13 only)	Calories
(R)	Grapefruit drink	83
(B)	Apricot cereal cubes (8)	114
(R)	Sausage patties (2)	223
(R)	Banana pudding	282
(R)	Fruit cocktail	87
Apricot cereal cubes (day 13 only)		789

Meal B		Calories
(R)	Beef and vegetables	98
(R)	Potato salad	143
(B)	Cheese sandwiches (6)	324
(B)	Strawberry cubes (6)	283
(R)	Orange drink	83
		931

Meal C		Calories
(R)	Orange-grapefruit drink	83
(R)	Tuna salad	214
(R)	Apricot pudding	150
(B)	Date fruitcake (4)	262
		709

Total Calories	2,429
Food Only Weight	521.12 gm

MENU II – DAYS 2,6, 10 & 14

Meal A		Calories
(R)	Grapefruit drink	83
(R)	Chicken and gravy	92
(B)	Beef sandwiches (6)	268
(R)	Applesauce	165
(B)	Peanut cubes (6)	297
		905

Meal B		Calories
(R)	Orange-grapefruit	83
(R)	Beef pot roast	119
(B)	Bacon & egg bites (6)	206
(R)	Chocolate pudding	307
		715

Meal C	(Days 2, 6, 10 only)	Calories
(R)	Potato soup	220
(R)	Shrimp cocktail	119
(B)	Date fruitcake (4)	262
(R)	Orange drink	83
		684

Total Calories	2,304
Food Only Weight	518.62 gm

MENU III – DAYS 3, 7 & 11

Meal A		Calories
(R)	Salmon salad	246
(R)	Green peas	81
(B)	Tosted bread cubes (8)	107
(B)	Gingerbread (6)	183
(R)	Cocoa	190
Toasted bread cubes (day 3 only)		807

Meal B		Calories
(R)	Grapefruit drink	83
(B)	Bacon squares (4)	90
(R)	Chicken & vegetables	75
(B)	Apricot cubes (6)	284
(B)	Pineapple fruitcake (6)	379
		911

Meal C		Calories
(B)	Cheese sandwich (6)	324
(R)	Butterscotch pudding	234
(R)	Orange drink	83
		711

Total Calories	2,429
Food Only Weight	515.06 gm

MENU IV – DAYS 4, 8 & 12

Meal A		Calories
(B)	Strawberry cereal cubes (8)	114
(B)	Bacon squares (4)	135
(R)	Ham and applesauce	127
(R)	Chocolate pudding	307
(R)	Orange drink	83
Strawberry cereal cubes (day 8 only)		766

Meal B		Calories
(R)	Beef and gravy	160
(R)	Corn chowder	252
(B)	Brownies (6)	241
(R)	Peaches	98
		751

Meal C		Calories
(B)	Coconut cubes (6)	310
(B)	Cinnamon toast (6)	99
(R)	Chicken salad	237
(R)	Applesauce	165
(R)	Grapefrut drink	83
		894

Total Calories	2,411
Food Only Weight	509.58 gm

(B) = Bite-sized food not requiring rehydration prior to ingestion. Usual serving consists of six bite-size pieces.
(R) = Rehydratable food, i.e., food which must be reconstituted prior to ingestion.

GEMINI 6 LAUNCH PREPARATIONS

The launching of Gemini 7 and the rapid turn-around for the Gemini 6 mission will be one of the most complex operations ever conducted by launch operations crews. A work schedule has been established for flight testing, checkout and launching of Gemini 6 nine days after Gemini 7.

The Gemini 6 spacecraft and launch vehicle were checked out thoroughly and counted down to some 42 minutes before liftoff on October 25. Since that time both the spacecraft and launch vehicle were placed in "bonded" storage under guard to insure that their mechanical and electrical integrity remains intact.

The Gemini 6 checkout will be the same as if a problem had occurred several days before the originally planned Gemini 6 flight and the spacecraft had to be de-mated from the launch vehicle. With the problem solved, the spacecraft would again be mated and a "compressed" checkout would take place in the days leading up to a launch, as most of the previous testing was still valid.

Once Gemini 7 has been launched, crews will be ready to erect the 6 launch vehicle and mate the spacecraft as soon as possible. The schedule calls for this to be completed some 24 hours after Gemini 7 liftoff.

Certain tests conducted during a normal mission will not have to be repeated for Gemini 6 because they will still remain valid (their validity will be checked, however). These include various calibrations of launch vehicle, spacecraft and blockhouse automatic ground support equipment; weight and balancing of the erector, certain spacecraft pre-mate and extensive spacecraft launch vehicle combined systems tests. Another departure from regular checkout procedure is that the Gemini 6 spacecraft will be fueled and the water supply and batteries will be installed before it gets to the launch pad.

The spacecraft and launch vehicle crews (McDonnell Aircraft is prime contractor for the spacecraft and Martin Company for the launch vehicle) will work on a three-shift, 24-hour schedule during the period between the launches.

The following is a general outline of the comprehensive work schedule:

- The Gemini 7 is scheduled for launch at 2:30 p.m. EST.
- As soon as possible an assessment of pad damage will be made. The blast damage has been very minimal in the past, requiring replacement primarily of some expendable wiring at the base of the pad. Various umbilical cables etc. will be checked to insure that they are operating. This work can be accomplished at the same time as the two launch vehicle stages are being erected and as the spacecraft is mated.
- During launch days plus one and two, preparations will be made for final spacecraft systems tests. The validity of the electrical interface between the spacecraft and launch vehicle will be verified on the third day as individual tests of the two continue.
- The final spacecraft systems tests are to be conducted on days three and four. During this time the previous verification of guidance between the spacecraft and launch vehicle is checked.
- Liquid oxygen for the spacecraft environmental control system will be loaded aboard during this time as preparations are made for the simulated flight, scheduled for launch day plus five. From this point, to Gemini 6 liftoff, the checkout will be generally the same as of any other Gemini flight.
- The simulated flight, which lasts some 10 to 12 hours, consists of three simulated launches – A mode II abort run (an abort occurs some 1:38 after liftoff), a switchover to secondary guidance during powered flight, and finally, a normal flight and insertion into orbit, during which various orbital exercises, reentry and recover tests are run. The prime pilots and their backups participate in these tests aboard the spacecraft at Launch Complex 19.
- Another "time saver" comes after the simulated flight when the various pyrotechnics aboard the spacecraft are thoroughly checked and connected. Since the pyrotechnics system was verified previously and left in a flight mode configuration, only a short test for verification will be required.

- The Gemini 6 pre-count (lasting some four hours on the third day before launch) and the mid-count (lasting some four hours on the second day before launch) will be very similar to regular Gemini procedures. During the precount significant portions of each spacecraft system are again checked.
- Final interface tests between the spacecraft and launch vehicle (guidance, abort procedures etc.) are conducted during the mid-count.
- The Gemini 7 and 6 final countdowns will be about the same – the spacecraft starting at about T-6 hours and the launch vehicle at T-4 hours. However, on Gemini 6, a hold will be declared at T-3 minute mark in the count to adjust the launch time to the planned rendezvous with Gemini 7. This hold will last 25 minutes. The Gemini 6 launch time is scheduled for 9:35 a.m. EST.

LAUNCH VEHICLE COUNTDOWN

Gemini 6

F-3 days ....Start pre-count
F-1 day ....Start mid-count
T-12 hours ....GLV propellant loading
T-390 minutes ....Complete propellant loading
T-300 minutes ....Begin terminal countdown
T-120 minutes ....Flight Crew to Complex 19
T-90 minutes ....Crew enters spacecraft
T-75 minutes ....Close spacecraft hatches
T-50 minutes ....White Room evacuation
T-35 minutes ....Begin erector lowering
T-15 minutes ....Spacecraft OAMS static firing
T-3 minutes ....25-minute hold
T-04 seconds ....GLV ignition
T-0 seconds ....Liftoff
T+2 minutes., 36 seconds ....Booster engine cutoff (BECO)
T+5:41 ....Second stage engine cutoff (SECO)
T+6:11 ....Spacecraft-launch vehicle separation

NOMINAL MISSION PLAN

Gemini 6

Gemini 6 is scheduled to be launched December 13 at about 9:34 am EST from Launch Complex 19 at Cape Kennedy, Florida. It will be launched into an elliptical orbit of 168 miles apogee and 100 miles perigee. Second stage booster yaw steering will be used to place the spacecraft into the same orbital plane as Gemini 7. Yaw steering provides up to 0.55 degree inclination increment change if needed. The spacecraft will trail Gemini 7 by 1208 miles at insertion.

Launch Windows (EST)

Nominal Day	9:34 a.m. to 10:21 a.m.	11:09 a.m. to 11:24 a.m.
N + 1	9:38 a.m. to 10:25 a.m.	
N + 2	8:07 a.m. to 8:54 a.m.	9:42 a.m. to 10:14 a.m.
N + 3	8:11 a.m. to 8:58 a.m.	
N + 4	6:59 a.m. to 7:25 a.m.	8:14 a.m. to 9:01 a.m.

Rendezvous is planned for the fourth orbit of Gemini 6, if liftoff is on time. During the first 35 minutes of each launch opportunity each 100 seconds delay in liftoff delays rendezvous by one spacecraft orbit. If liftoff time occurs beyond 300 seconds, rendezvous will not be attempted until the beginning of the second day when better tracking coverage is available.

Should liftoff occur during the last 12 minutes of the maximum 47 minute window, a different intermediate sequence of maneuvers will be initiated to narrow the catch-up distance between the spacecraft. In this case, engine cutoff occurs earlier to reduce velocity by 50 feet per second. This causes the spacecraft to be inserted into a lower orbit than planned, with a perigee of about 100 miles and apogee of 138 miles. In this orbit the Gemini 6 catch-up rate will be increased due to the greater difference in altitude between the two spacecraft.

Varying insertion velocity as described above has the effect of widening the launch window.

Following a successful, on-time liftoff and insertion the uncertainties of the effect of drag on the spacecraft during its initial orbit may require a one foot per second posigrade burn at first perigee to raise apogee. In the event of small insertion dispersions, the magnitude of this maneuver may vary but the resulting apogee will be 168 miles.

Near the second apogee a posigrade burn will add 53 feet per second to raise perigee to about 134 miles. This reduces the catch-up rate from 6.7 degrees to 4.5 degrees per orbit and will provide the proper phase relationship between the two spacecraft for circularization at third apogee.

Should the two spacecraft be in different planes, a plane adjustment will be made by Gemini 6 at the common node (where the two spacecraft orbits intersect) following the second apogee posigrade burn.

At the third Gemini 6 spacecraft apogee, a posigrade burn of 53 feet per second will be made to circularize the orbit at 146 miles.

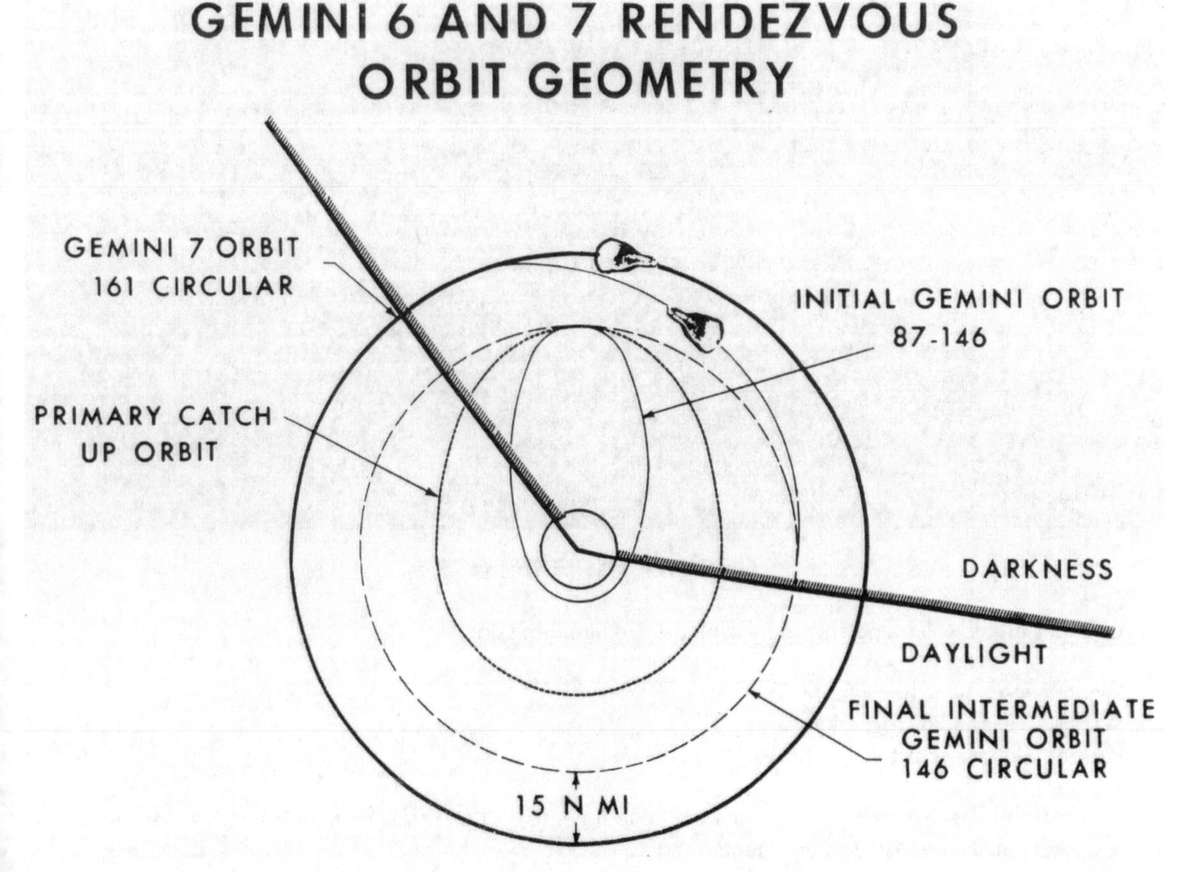

Gemini 6 will then be trailing Gemini 7 by about 184 miles. This is within range of the onboard radar and lock-on should have occurred.

A 32 feet per second posigrade burn will be made at terminal phase initiation along the line of sight to Gemini 7. This will be at a ground elapsed time of about 5 hours, 15 minutes – about one minute after entering darkness. The range between the spacecraft at this time is expected to be about 39 miles.

Approximately 33 minutes following terminal phase initiation a posigrade velocity of 43 feet per second will be applied to Gemini 6. This places the two spacecraft into the same orbit and rendezvous will have been accomplished.

Should there be computer, platform or radar failure, the mission can still proceed using ground data and radar-optical or optical rendezvous modes.

Retrofire will occur at a ground elapsed time of 46 hours and 10 minutes, during the 29th revolution. Landing will be in the West Atlantic recovery area.

GEMINI 6 EXPERIMENTS

Three experiments will be performed during the Gemini 6 mission:

1. Synoptic Weather Photography
2. Synoptic Terrain Photography
3. Radiation in spacecraft

The photography experiments are repeats of those flown on all previous Gemini flights. A description of these appears in the Gemini 7 experiments section.

The radiation experiment is designed to measure radiation levels and distribution inside the spacecraft. Seven sensors are located throughout the spacecraft. One is shielded to simulate the amount of radiation the crew members are receiving beneath their skin. The shield will be removed as the spacecraft passes through the South Atlantic anomaly, the area where the radiation belt dips closest to the Earth's surface.

This experiment was also flown on Gemini 4.

MANNED SPACE FLIGHT NETWORK
GEMINI 7 AND 6 MISSIONS

The Manned Space Flight Network consists of NASA and Department of Defense facilities.

The Mission Control Center in Houston, Texas (MCC-H) will control the entire Gemini 7-6 mission. As on Gemini missions 4 and 5, Houston's Real-Time Computer Complex (RTCC) (a key element of the MCC-H) will serve as the mission computing center.

For Gemini 7-6 the network will provide:

(1) Tracking and telemetry data during launch and orbital phases from both the Gemini 7 and the Gemini 6 spacecraft for position determination and systems operation.
(2) Capability for transmission and verification of ground commands to either or both spacecraft. These commands, generated at the mission control center, update the spacecraft computer to provide current information for time-of-retrofire determination and reentry calculations and displays.

Prime Computing Support

Immediate computing support will be provided from launch through impact by the RTCC at the Manned Spacecraft Center. During the launch and insertion phase, the RTCC will receive high-speed radar data from Bermuda and radar and MISTRAM (Missile Tracking and Measurement System) data from the Air Force Eastern Test Range (AFETR) radars via the Cape Kennedy-Houston GLDS (Gemini Launch Data System).

Other Computer Support

NASA's Goddard Space Flight Center (GSFC) real-time computing support for Gemini 7-6 includes the processing of skin tracking information obtained from the second stages of both launch vehicles and the computation of their predicted impact points. Additionally, the GSFC RTCC will generate skin track (radar echo bounce) space position predictions for the Manned Space Flight Network and the Department of Defense for their use in the event of spacecraft beacon loss or powered-down flight.

Network Readiness

Computers at the GSFC will certify the worldwide network's readiness to support Gemini 7-6 through a system-by-system TLM, CMD, RDR, station-by-station, computer-programmed checkout method called CADFISS (Computation and Data Flow Integrated Subsystem Tests). Checkout of network facilities also will be performed by the GSFC during post-launch periods when the spacecraft are not electronically "visible" by some stations and continue until the vehicles are again within acquisition range.

Data Flow Tests (DFT's) from the worldwide network to the Manned Spacecraft Center's Real-Time Computing Complex will be conducted from the Manned Spacecraft Center under the direction of the CADFISS Test Director.

TRACKING TWO MANNED SPACECRAFT

For Gemini 7-6, various combinations of spacecraft tracking and data acquisition assignments will be accomplished according to individual station capability.

Both Gemini spacecraft are equipped with C-Band beacon systems that aid station radars of the Manned Space Flight Network in pinpointing precise space position of each vehicle. In order that these radiated signals be readily distinguishable by the ground systems, their identifying codes have been altered slightly for precise recognition.

Through the detection of the spacecraft beacon, ground trackers pinpoint each spacecraft with an accuracy equivalent to a 22 bullet hitting a twenty-five cent piece at a distance of one mile. Electronically coordinated (slaved) telemetry receiving and radio command antenna systems acquire data from and send instructions to each spacecraft based in part on the space position information provided by the beacon tracking radars.

After Gemini 6 spacecraft insertion into orbit, stations in the Manned Space Flight Network will, for the first time, simultaneously track and acquire information from two orbiting manned spacecraft.

While both spacecraft are in orbital flight the on-site flight data summary computer (UNIVAC 1218) called TOMCAT (Telemetry On-Line Monitoring, Compression and Transmission) processes Gemini 6 or 7 (whichever spacecraft has been designated "prime") spacecraft data.

Upon termination of the Gemini 6 mission, all network station systems will revert to previous Gemini 7 operating modes for the remainder of the mission.

ORBITS – REVOLUTIONS

The spacecraft's course is measured in revolutions around the Earth. A revolution is completed each time the spacecraft passes over 80 degrees west longitude, or at Gemini altitudes about once every 96 minutes.

Orbits are space referenced and in Gemini take about 90 minutes.

CREW TRAINING BACKGROUND

In addition to the extensive general training received prior to flight assignment the following preparations have or will be accomplished prior to launch:

1. Launch abort training in the Gemini Mission Simulator and the Dynamic Crew Procedures Simulator.
2. Egress and recovery activities using a spacecraft boilerplate model and actual recovery equipment and personnel. Pad emergency egress training using elevator and slide wire.
3. Celestial pattern recognition in the Moorehead Planetarium, Chapel Hill, North Carolina.
4. Parachute descent training over water using a towed parachute technique.
5. Zero gravity training in KC-135 aircraft.
6. Suit, seat and harness fittings.
7. Training sessions for each crew member on the Gemini translation and docking simulator.
8. Detailed systems briefing; detailed experiment briefings; flight plans and mission rules reviews.
9. Participation in mockup reviews, systems review, subsystem tests and spacecraft acceptance review.

During final preparation for flight, the crew participates in network launch abort simulations, joint combined systems test and the final simulated flight tests. At T-two days, the major flight crew medical examinations will be administered to confirm readiness for flight and obtain data for comparison with post-flight medical examination results.

IMMEDIATE PRE-FLIGHT CREW ACTIVITIES

T-7 hours	Back-up flight crew reports to the 100-foot level of the White Room to participate in final flight preparations.
T-5 hours	Pilot's ready room, 100-foot level of White Room and crew quarters manned and made ready for prime crew.
T-4 hours, 30 minutes	Primary crew awakened
T-4 hours	Medical examination
T-3 hours, 40 minutes	Breakfast
T-3 hours, 15 minutes	Crew leaves quarters
T-3 hours, 5 minutes	Crew arrives at ready room on Pad 16

During the next hour, the biomedical sensors are placed, underwear and signal conditioners are donned, flight suits minus helmets and gloves are put on and blood pressure is checked. The helmets and gloves are then attached and communications and oral temperatures systems are checked.

T-2 hours, 15 minutes	Purging of suit begins
T-2 hours, 4 minutes	Crew leaves ready room
T- 2 hours	Crew arrives at 100-foot level
T-1 hour, 30 minutes	Crew enters spacecraft

From entry until ignition, the crew participates in or monitors systems checks and preparations.

Flight Activities

At ignition the crew begins the primary launch phase task of assessing system status and detecting abort situations. Thirty seconds after SECO, the command pilot initiates forward thrusting and the pilot actuates spacecraft separation and selects rate command attitude control. Ground computations of insertion velocity corrections are received and velocity adjustments are made by forward or aft thrusting. After successful insertion and completion of the insertion check list, the detailed flight plan is begun.

CREW SAFETY

Every Gemini system affecting crew safety has a redundant (Back-up) feature. The Malfunction Detection System aboard the launch vehicle monitors subsystem performance and warns the crew of a potentially catastrophic malfunction in time for escape.

There are three modes of escape:

MODE I — Ejection seats, and personal parachute, used at ground level and during first 50 seconds of powered flight, or during descent after reentry.

MODE II (Delayed) Retrorockets used between 50 and 100 seconds, allowing crew to salvo fire all four solid retrorockets five seconds after engine shutdown is commanded.

MODE III — Normal separation from launch vehicle, using OAMS thrusters, then making normal reentry, using computer.

Except for Mode I, spacecraft separates from Gemini Launch Vehicle, turns blunt-end forward, then completes reentry and landing with crew aboard.

Survival Package

Survival gear, mounted on each ejection seat and attached to the astronaut's parachute harnesses by nylon line, weighs 23 pounds.

Each astronaut has:

- 3.5 pounds of drinking water.
- Machete.
- One-man life raft, 5½ by 3 feet, with CO_2 bottle for inflation, sea anchor, dye markers, nylon sun bonnet.
- Survival light (strobe), with flashlight, signal mirror, compass, sewing kit, 14 feet of nylon line, cotton balls and striker, halazone tablets, a whistle, and batteries for power.
- Survival radio, with homing beacon and voice reception.
- Sunglasses.

- Desalter kit, with brickettes enough to desalt eight pints of sea water.
- Medical kit, containing stimulant, pain, motion sickness and antibiotic tablets and aspirin, plus injectors for pain and motion sickness.

GEMINI 6 SUIT

The pressure suit worn by the crew of Gemini 6 is identical to that worn by the Gemini 5 crew. It is not suitable for extravehicular activity.

It has five layers:
1. White cotton constant wear undergarment with pockets to hold biomedical instrumentation equipment.
2. Blue nylon comfort layer.
3. Black neoprene-coated nylon pressure garment.
4. Restraint layer of dacron and teflon link net to restrain pressure garment and maintain its shape.
5. White HT-1 nylon outer layer to protect against wear and solar reflectance.

The suit is a full pressure garment, including a helmet with mechanically sealed visor. Oxygen is furnished by the environmental control system. Gaseous oxygen is provided to the suit through a "suit loop" to cool the astronaut and provide him with a breathable atmosphere of 100 percent oxygen. Oxygen in the cabin maintains 5.1 pounds per square inch (psi) pressure. The suit, if cabin pressure fails, is pressurized to 3.5 psi (+.4, -0).

GEMINI 7 SUIT

A new lightweight suit has been developed for long duration space flights. It will be worn for the first time by the Gemini 7 crew. It is an intravehicular suit designed to give maximum mobility when depressurized.

It has two layers:
1. The inner layer is the pressure restraining neoprene-coated nylon bladder.
2. The outer layer is six ounce HT-1 nylon.

It is a full pressure suit and weighs 16 pounds, including an aviator's crash helmet which is worn under the soft helmet. The suit can be completely taken off during flight or can be worn in a partially doffed mode in which gloves and boots are removed and the helmet is unzipped at the neck and rolled back to form a headrest.

Emergency time to don the suit from a partially doffed mode is about 35 seconds. When the suit is totally doffed it takes from five to ten minutes to don it.

MEDICAL CHECKS

At least one medical check a day will be made by each crew member. Performed over a convenient ground station, a check will consist of: Oral temperature, blood pressure measurement, food and water intake evaluation.

BODY WASTE DISPOSAL

Two separate systems are used for collection of body wastes.

A plastic bag with an adhesive lip to provide secure attachment to the body is used for the collection of feces. It contains a germicide which prevents formation of bacteria and gas. Soiled items, toilet tissues and a wet towel, are placed in the bag following use. The adhesive lip is then used to form a liquid seal and the bag is rolled and stowed in the empty food container spaces and brought back to earth for analysis.

Urine is collected into a horn-shaped receptacle with a self adjusting opening. The receptacle is connected by a hose to a pump device which either transfers the liquid to the evaporator or dumps it overboard. The system is much like the relief tube used in military fighter planes.

FOOD

Number of Meals – Three per day per astronaut.

Type – Bite-sized and rehydratable. Water is placed in rehydratables with special gun. Bite-sized items need no rehydration.

Storage – Meals individually wrapped in aluminum foil and polyethylene, polyamide laminate. First day meals stored in compartment beside knees of each crewman. Succeeding days meals in right aft food compartment.

The water intake of each astronaut will be carefully measured. A mechanical measuring system is an integral part of the water gun. It consists of a neoprene bellows housed in a small metal cylinder mounted at base of gun. The bellows holds one-half ounce of water. When plunger of gun is depressed, a spring pushes water out of bellows and through gun. A counter in right side of gun registers number of times bellows is activated. Each crewman will record how much he drinks by noting numbers at beginning and end of use of gun.

WEATHER REQUIREMENTS

The following are guidelines only. Conditions along the ground track will be evaluated prior to and during the mission.

Launch Area
Surface Winds – 18 knots with gusts to 25 knots
Ceiling – 5,000 feet cloud base
Visibility – Six miles
Wave Height – Five feet maximum

Planned Landing Areas
Surface Winds – 30 knots maximum
Ceiling – 1,500 feet cloud base
Visibility – Six miles
Wave Height – Eight feet maximum

Contingency Landing Areas
Flight director will make decision based upon conditions at the time.

Pararescue
Surface Winds – 25 knots maximum
Ceiling – 1,000 feet cloud base
Visibility – Target visible
Waves – Five feet maximum; swells 10 or 11 feet maximum

PLANNED AND CONTINGENCY LANDING AREAS

There are two types of landing areas for Gemini spacecraft, planned and contingency. Planned areas are those where recovery forces are pre-positioned to recover spacecraft and crew within a short time. All other areas under the orbital track are contingency areas, requiring special search and rescue techniques and a longer recovery period.

Planned Landing Areas

PRIMARY	Landing in the West Atlantic where the primary recovery vessel, an aircraft carrier, is pre-positioned.
SECONDARY	Landing in East Atlantic, West Pacific and Mid-Pacific areas where ships are deployed.
LAUNCH SITE	Landing in the event of off-the-pad abort for abort during early phase of flight, includes an area about 41 miles seaward from Cape Kennedy, 3 miles toward Banana River from Complex 19.
LAUNCH ABORT	Landing in the event of abort during powered flight, extending from 41 miles at sea from Cape Kennedy to west coast of Africa.

Contingency Landing Areas

All the area beneath the spacecraft's ground track except those designated Planned Landing Areas are Contingency Landing Areas, requiring aircraft and pararescue support for recovery within a period of 18 hours from splashdown.

Recovery forces will be provided by the military services, and during mission time will be under the operational control of the Department of Defense Manager for Manned Space Flight Support Operations.

GEMINI SPACECRAFT

The Gemini spacecraft is conical, 18 feet, 5 inches long, 10 feet in diameter at its base and 39 inches in diameter at the top. Its two major sections are the reentry module and the adapter section.

Reentry Module

The reentry module is 11 feet high and 7½ feet in diameter at its base. It has three main sections: (1) rendezvous and recovery (R&R), reentry control (RCS), and (3) cabin.

Rendezvous and recovery section is the forward (small) end of the spacecraft, containing drogue, pilot and main parachutes and radar.

Reentry control section between R&R and cabin sections contains fuel and oxidizer tanks, valves, tubing and two rings of eight attitude control thrusters each for control during reentry. A parachute adapter assembly is included for main parachute attachment.

Cabin section between RCS and adapter section, houses the crew seated side-by-side, their instruments and controls. Above each seat is a hatch. Crew compartment is a pressurized titanium hull. Equipment not requiring pressurized environment is located between pressure hull and outer beryllium shell which is corrugated and shingled to provide aerodynamic and heat protection. Dish-shaped heat shield forms the large end of cabin section.

Adapter Section

The adapter section is 7½ feet high and 10 feet in diameter at its base, containing retrograde and equipment sections.

Retrograde section contains four solid retrograde rockets and part of the radiator for the cooling system.

Equipment section contains electrical power source systems, fuel for the orbit attitude and maneuver system (OAMS), primary oxygen for the environmental control system (ECS). It also serves as a radiator for the cooling system, also contained in the equipment section.

RCS FUNCTION

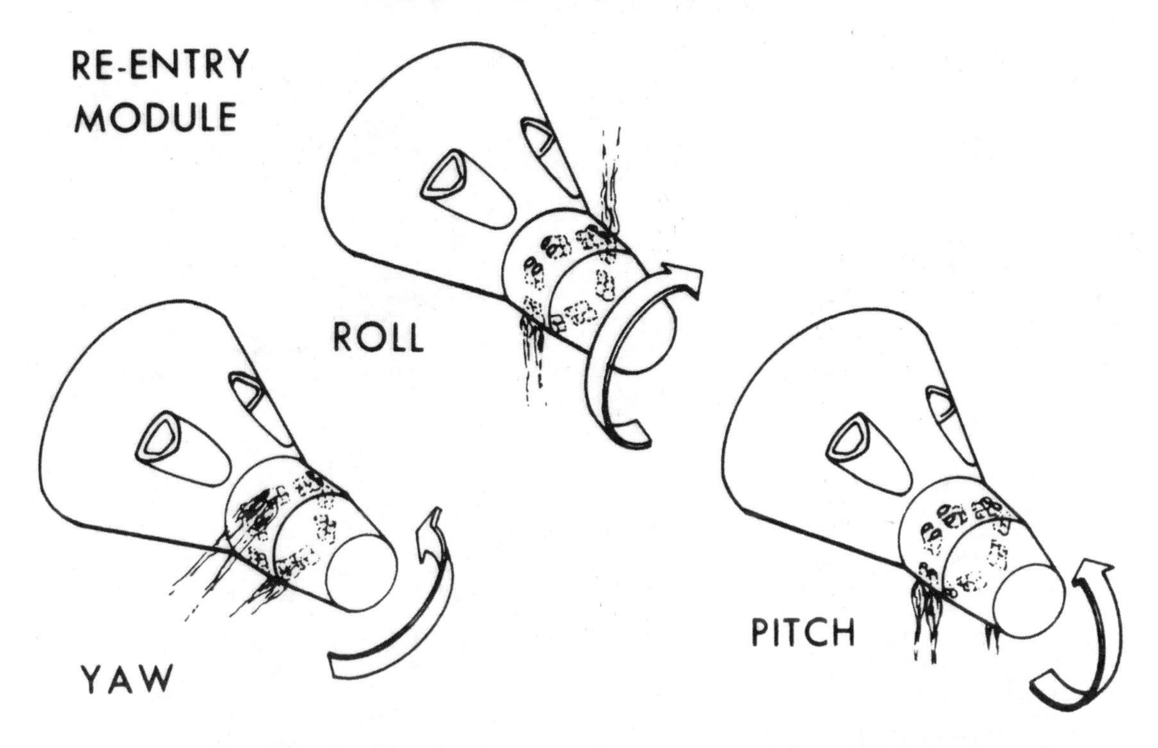

SPACECRAFT RESPONSES TO ORBIT ATTITUDE CONTROL THRUST

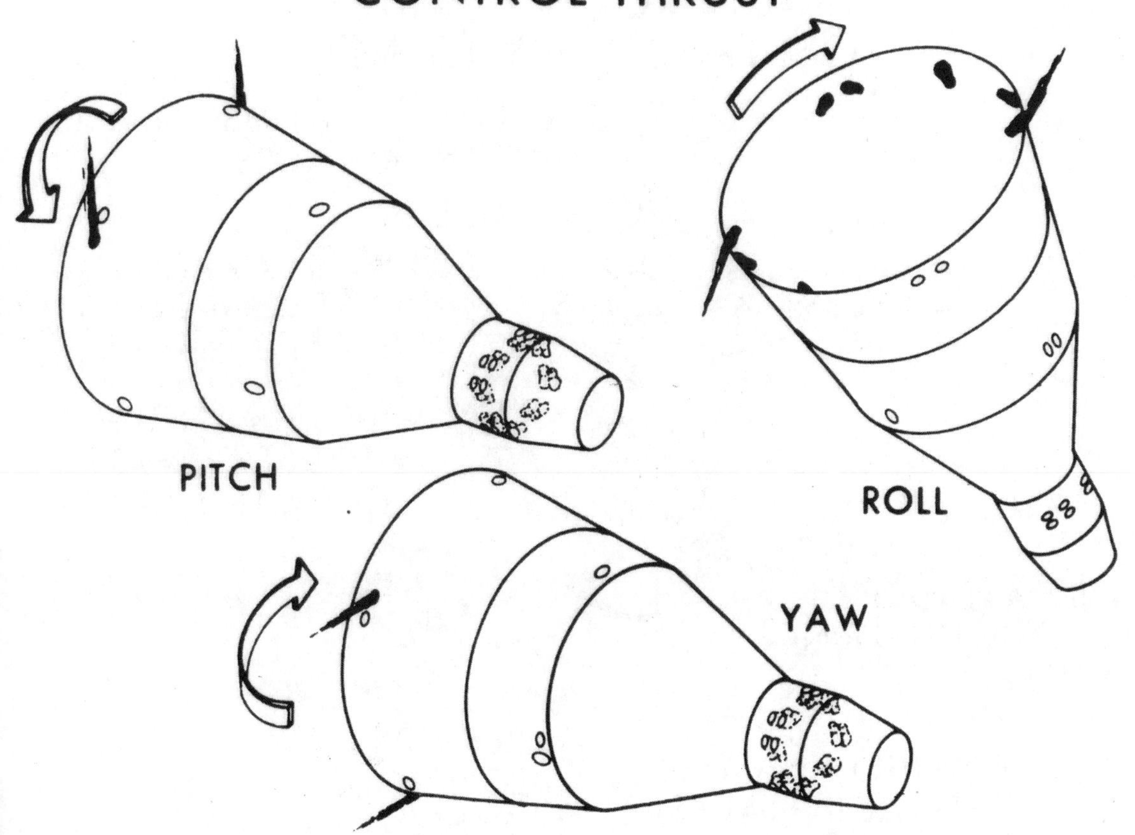

MANEUVERING CONTROL

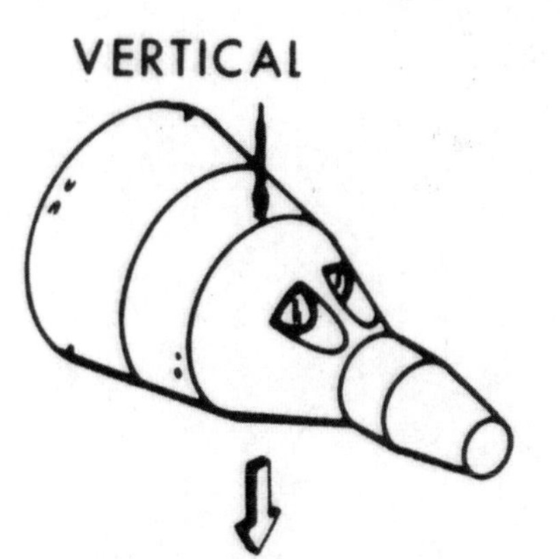

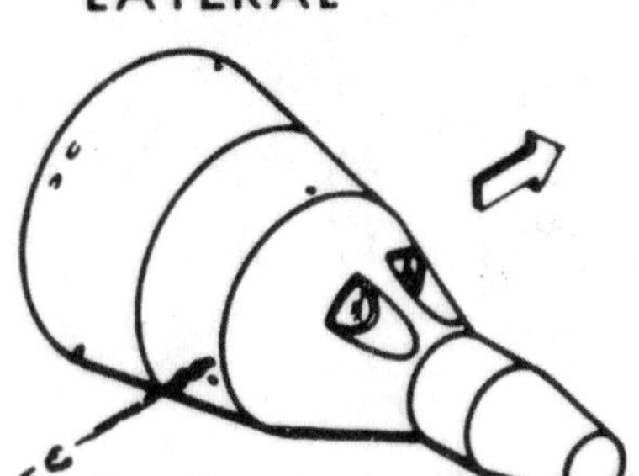

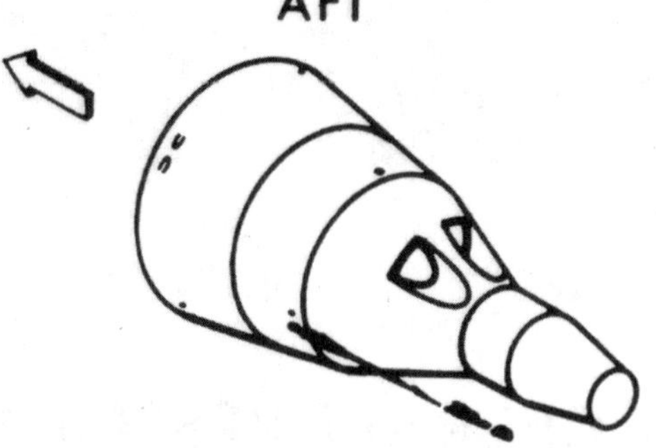

LIQUID ROCKET SYSTEMS GENERAL ARRANGEMENT

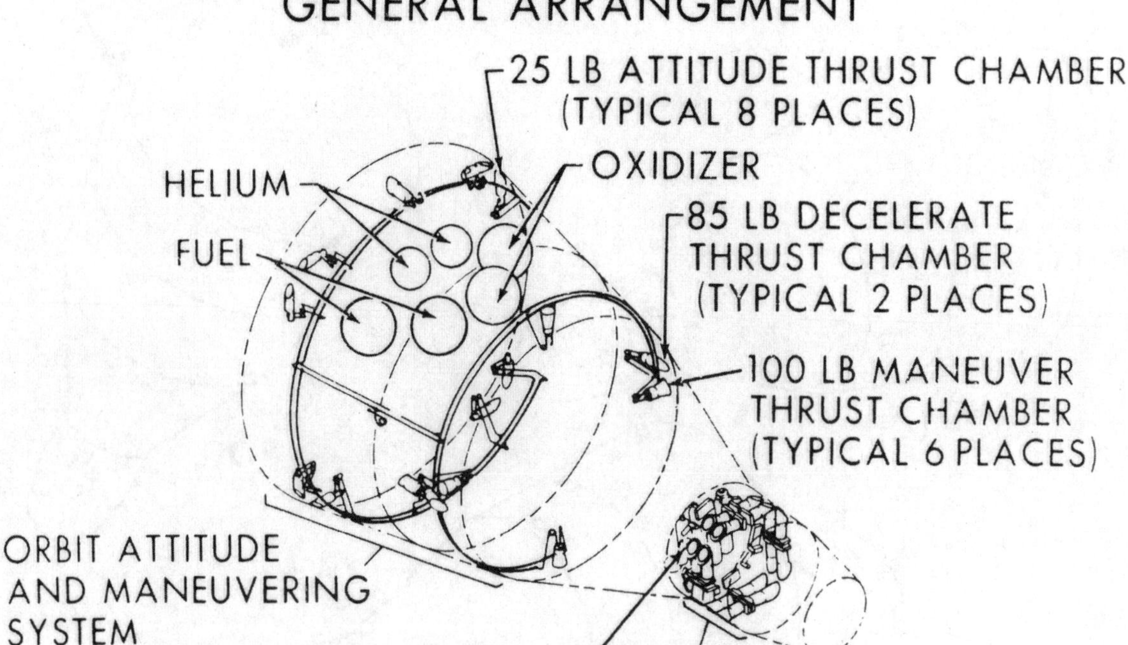

NOTE: The equipment section is jettisoned immediately before retrorockets are fired for reentry. The retrograde section is jettisoned after retros are fired.

PROPELLANT

Gemini-7 – 423 pounds

Gemini-6 – 669 pounds

GEMINI-7 SPACECRAFT MODIFICATIONS

The following modifications have been made to the Gemini-7 spacecraft to support Gemini-6 rendezvous mission:

1. A transponder to receive and transmit signals from the Gemini-6 rendezvous radar system has been installed in the nose of the spacecraft.
2. Two acquisition lights have been placed on the adapter section 180 degrees apart. These are the same lights designed for the Agena target vehicle. They flash about 80 times per minute and can be seen for approximately 23 miles.

RENDEZVOUS RADAR

Gemini-6

Purpose – Enables crew to measure range, range rate, and bearing angle to Gemini-7. Supplies data to Inertial Guidance System computer so crew can determine maneuvers necessary for rendezvous.

Operation – Transponder on Gemini 7 receives radar impulses and returns them to Gemini 6 at a specific frequency and pulse width. Radar accepts only signals processed by transponder.

Location – small end of spacecraft on forward face of rendezvous and recovery section.

Size – less than two cubic feet.

Weight – less than 50 pounds.

Power Requirement – less than 80 watts.

ELECTRICAL POWER SYSTEMS

Gemini-7

The fuel cell power subsystem includes two 68-pound pressurized fuel cell sections, each containing three fuel cell stacks of 32 series-connected cells. Operating together, these sections produce up to two kilowatts of DC power at peak load.

Four conventional silver zinc batteries provide backup power to the fuel cells during launch and are primary power for reentry, landing and post-landing. Three additional batteries are isolated electrically to activate pyrotechnics aboard the spacecraft. (The four main batteries can also be brought on line for this purpose if necessary.)

Besides its two cylindrical sections, the fuel cell battery subsystem includes a reactant supply of hydrogen and oxygen, stored at supercritical pressures and cryogenic temperatures.

Energy is produced in the fuel cell by forcing the reactants into the stacks where they are chemically changed by an electrolyte of polymer plastic and a catalyst of platinum. Resultant electrons and ions combine with oxygen to form electricity, heat and water. This chemical reaction will theoretically continue as long as fuel and oxidant are supplied. Electricity is used for power, heat is rejected by the spacecraft coolant system, and water is diverted into the spacecraft drinking supply tanks where it is separated from the crew's drinking supply by a bladder and used as pressurant to supply drinking water.

MAIN FUEL CELL BATTERY AND BATTERY CONSTRUCTION (BELOW)

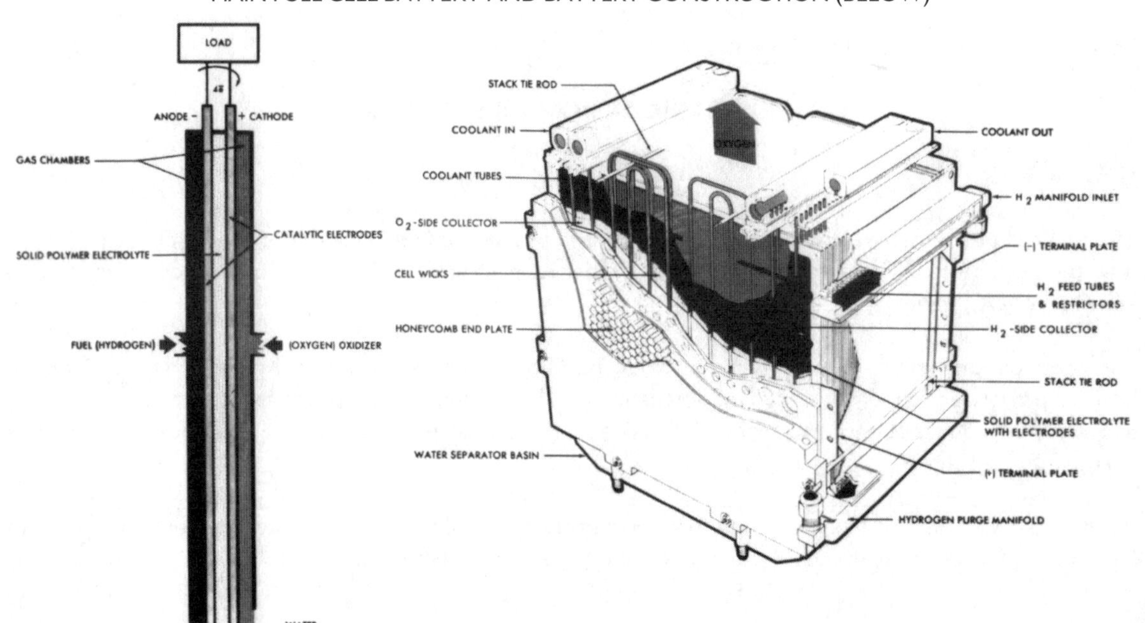

Gemini-6

Gemini-6 spacecraft carries 10 batteries. Included in these are:

Adapter Batteries: three 400-amp/hour units, housed in the adapter section. Primary power source.
Main Batteries: four 45-amp/hour units in the reentry section for power prior to and during reentry.
Squib Batteries: three 15-amp/hour units in the reentry section, used to trigger explosive squibs.

GEMINI LAUNCH VEHICLE

The Gemini Launch Vehicle (GLV) is a modified U.S. Air Force Titan II intercontinental ballistic missile consisting of two stages.

GLV dimensions are:

	First Stage	Second Stage
HEIGHT	63 feet	27 feet
DIAMETER	10 feet	10 feet
THRUST	430,000 pounds (two engines)	100,000 pounds (one engine)
FUEL	50-50 blend of monomethyl hydrazine and unsymmetrical-dimethyl hydrazine	
OXIDIZER	Nitrogen tetroxide (Fuel is hypergolic, ignites spontaneously upon contact with oxidizer).	

Overall height of launch vehicle and spacecraft is 109 feet. Combined weight is about 340,000 pounds.

Modifications to Titan II for use as the Gemini Launch Vehicle include:

1. Malfunction detection system added to detect and transmit booster performance information to the crew.
2. Backup flight control system added to provide a secondary system if primary system fails.
3. Radio guidance substituted for inertial guidance.
4. Retro and vernier rockets deleted.
5. New second stage equipment truss added.
6. New second stage forward oxidizer skirt assembly added.
7. Trajectory tracking requirements simplified.
8. Electrical, hydraulic and instrument systems modified.

Gemini Launch Vehicle program management for NASA is under the direction of the Space Systems Division of the Air Force Systems Command.

CREW BIOGRAPHIES

FRANK BORMAN, Gemini 7 command pilot

BORN: Gary, Indiana, March 14, 1928
HEIGHT: 5 feet, 10 inches; WEIGHT: 163 lbs.; Blonde hair, blue eyes
EDUCATION: Bachelor of Science degree, United States Military Academy, 1950; Master of Science degree in aeronautical engineering, California Institute of Technology, 1957.
MARITAL STATUS: Married to the former Susan Bugbee of Tucson, Arizona.
CHILDREN: Frederick, October 4, 1951; Edwin, July 20, 1953
EXPERIENCE: Upon graduation from West Point, Borman, now an Air Force Major, chose an Air Force career and received his pilot training at Williams Air Force Base, California.

From 1951 to 1956 he served with fighter squadrons in the United States and in the Philippines and was an instructor at the Air Force Fighter Weapons School.

From 1957 to 1960 he was an instructor of thermodynamics and fluid mechanics at the U.S. Military Academy.

He was graduated from the USAF Aerospace Research Pilots School in 1960 and later served there as an instructor. In this capacity he prepared and delivered academic lectures and simulator briefings, and flight test briefings on the theory and practice of spacecraft testing.

Borman has logged more than 4,400 hours flying time, including more than 3,600 hours in jet aircraft.

CURRENT ASSIGNMENT: Borman was one of the nine astronauts named by NASA in September 1962.

Borman is the son of Mr. and Mrs. Edwin Borman, Phoenix, Arizona.

JAMES A. (for Arthur) LOVELL, Jr., Gemini 7 pilot

BORN: Cleveland, Ohio, March 25, 1928
HEIGHT: 6 feet; WEIGHT: 165 lbs.; Blond hair, blue eyes
EDUCATION: Bachelor of Science degree from the United States Naval Academy, 1952; attended University of Wisconsin 1946-1948.
MARITAL STATUS: Married to the former Marilyn Gerlach of Milwaukee
CHILDREN: Barbara Lynn, October 13, 1953; James A., February 15, 1955; Susan Kay, July 14, 1958
EXPERIENCE: Lovell, a Navy Lieutenant Commander, received flight training following his graduation from Annapolis.

He served in a number of Naval aviator assignments including a three-year tour as a test pilot at the Naval Air Test Center at Patuxent River, Maryland. His duties there included service as program manager for the F4H Weapon System Evaluation.

Lovell was graduated from the Aviation Safety School of the University of Southern California.

He served as flight instructor and safety officer with Fighter Squadron 101 at the Naval Air Station at Oceana, Virginia.

Lovell has logged 3,000 hours flying time, including more than 2,000 hours in jet aircraft.

CURRENT ASSIGNMENT: Lovell was selected as an astronaut by NASA in September 1962. In addition to participating in the overall astronaut training program, he has been assigned special duties monitoring design and development of recovery and including crew life support systems and developing techniques for lunar and Earth landings and recovery.

Lovell is the son of Mr. and Mrs. James A. Lovell, Sr., Edgewater Beach, Florida.

EDWARD H. (for Higgins) WHITE II, Gemini 7 backup command pilot

BORN: San Antonio, Texas, November 14, 1930
HEIGHT: 6 feet; WEIGHT: 171 lbs.; Brown hair, brown eyes
EDUCATION: Bachelor of Science degree from United States Military Academy, 1952, Master of Science degree in aeronautical engineering, University of Michigan, 1959
MARITAL STATUS: Married to the former Patricia Eileen Finegan of Washington, D.C.
CHILDREN: Edward, May 15, 1953; Bonnie Lynn, September 15, 1950
PROFESSIONAL ORGANIZATIONS: Associate member of Institute of Aerospace Sciences; member of Sigma Delta Psi, athletic honorary; and member of Tau Beta Pi, engineering honorary
EXPERIENCE: White, an Air Force Major, received flight training in Florida and Texas, following his graduation from West Point. He spent 32 years in Germany with a fighter squadron, flying F-86's and F-100's.

He attended the Air Force Test Pilot School at Edwards Air Force Base, California, in 1959.

White was later assigned to Wright-Patterson Air Force Base, Ohio, as an experimental test pilot with the Aeronautical Systems Division. In this assignment he made flight tests for research and weapons systems development, wrote technical engineering reports, and made recommendations for improvement in aircraft design and construction.

He has logged more than 3,600 hours flying time, including more than 2,200 hours in jet aircraft.

CURRENT ASSIGNMENT: White is a member of the astronaut team selected by NASA in September 1962. He was assigned as the pilot for the second manned Gemini mission which flew for four days (June 3-7, 1965). White was the first U.S. astronaut to take part in extravehicular activities. He was outside the Gemini 4 spacecraft for 22 minutes and was the first human to use a personal propulsion unit for maneuvering in space.

White is the son of Major General and Mrs. Edward H. White, St. Petersburg, Florida.

MICHAEL COLLINS, Gemini 7 backup pilot

BORN: Rome Italy, October 31, 1930
HEIGHT: 5 feet, 10½ inches; WEIGHT: 163 lbs.; Brown hair, brown eyes
EDUCATION: Bachelor of Science degree from United States Military Academy
MARITAL STATUS: Married to the former Patricia M. Finnegan of Boston, Massachusetts
CHILDREN: Kathleen, May 6, 1959; Ann S., October 31, 1961; Michael L., February 23, 1963
EXPERIENCE: Collins, an Air Force Major, chose an Air Force career following graduation from West Point.

He served as an experimental flight test officer at the Air Force Flight Test Center, Edwards Air Force Base, California. In that capacity, he tested performance and stability and control characteristics of Air Force aircraft, primarily jet fighters.

He has logged more than 3,000 hours flying time, including more than 2,700 hours in jet aircraft. He is a member of the Society of Experimental Test Pilots.

Collins was one of the third group of astronauts selected by NASA in October 1963.

He is the son of the late Major General James L. Collins and Mrs. James L. Collins of Washington, D.C.

WALTER M. (for Marty) SCHIRRA, Jr., Gemini 6 command pilot

BORN: Hackensack, N.J., March 12, 1923
HEIGHT: 5 feet, 10 inches; WEIGHT: 170 lbs.; Brown hair, brown eyes
EDUCATION: Bachelor of Science degree, United States Naval Academy, 1945
MARITAL STATUS: Married to the former Josephine Fraser of Seattle, Washington.
CHILDREN: Walter M. III, June 23, 1950; Suzanne, September 29, 1957
EXPERIENCE: Schirra, a Navy Captain, received flight training at Pensacola Naval Air Station, Florida. As an exchange pilot with the United States Air Force, 154th Fighter Bomber Squadron, he flew 90 combat missions in F-84E aircraft in Korea and downed one MIG with another probable. He received the Distinguished Flying Cross and two Air Medals for his Korean service.

He took part in the development of the Sidewinder missile at the Naval Ordnance Training Station, China Lake, California. Schirra was project pilot for the F7U3 Cutlass and instructor pilot for the Cutlass and the FJ3 Fury.

Schirra flew F3H-2N Demons as operations officer of the 124th Fighter Squadron onboard the Carrier Lexington in the Pacific.

He attended the Naval Air Safety Officer School at the University of Southern California, and completed test pilot training at the Naval Air Center, Patuxent River, Maryland. He was later assigned at Patuxent in suitability development work on the F4H.

He has more than 3,800 hours flying time, including more than 2,700 hours in jet aircraft.

Schirra was one of the seven Mercury astronauts named in April 1959.

On October 3, 1962, Schirra flew a six-orbit mission in his "Sigma 7" spacecraft. The flight lasted nine hours and 13 minutes from liftoff through landing and he attained a velocity of 17,557 miles (28,200 kilometers) per hour, a maximum orbital altitude of 175 statute miles (281 kilometers) and a total range of almost 144,000 statute miles (231,700 kilometers). The impact point was in the Pacific Ocean, about 275 miles (443 kilometers) northeast of Midway Island. He was awarded the NASA Distinguished Service Medal for his flight. He was the backup command pilot for the Gemini 3 mission. Schirra is the son of Mr. and Mrs. Walter M. Schirra, Sr., San Diego, California.

THOMAS P. (for Patten) STAFFORD, Gemini 6 pilot

BORN: Weatherford, Oklahoma, September 17, 1930
HEIGHT: 6 feet; WEIGHT: 175 lbs.; Black hair, blue eyes
EDUCATION: Bachelor of Science degree from United States Naval Academy, 1951
MARITAL STATUS: Married to the Former Faye L. Shoemaker of Weatherford, Oklahoma.
CHILDREN: Dianne, July 2, 1954; Karin, August 28, 1957
EXPERIENCE: Stafford, an Air Force Major, was commissioned in the United States Air Force upon graduation from the U.S. Naval Academy at Annapolis. Following his flight training, he flew fighter interceptor aircraft in the United States and Germany, and later attended the United States Air Force Experimental Flight Test School at Edwards Air Force Base, California.

He served as Chief of the Performance Branch, USAF Aerospace Research Pilot School at Edwards. In this assignment he was responsible for supervision and administration of the flying curriculum for student test pilots. He established basic text books and participated in and directed the writing of flight test manuals for use by the staff and students.

Stafford is co-author of the Pilot's Handbook for Performance Flight Testing and Aerodynamic Handbook for Performance Flight Testing.

He has logged more than 4,300 hours flying time, including more than 3,600 hours in

jet aircraft.
Stafford was one of the nine astronauts named by NASA in September 1962. He was the backup pilot for Gemini 3. Stafford is the son of Mrs. Mary E. Stafford and the late Dr. Thomas S. Stafford, Weatherford, Oklahoma.

VIRGIL I. (for Ivan) "GUS" GRISSOM, Gemini 6 backup command pilot

BORN: Mitchell, Indiana, April 3, 1926
HEIGHT: 5 feet 7 inches; WEIGHT: 150 lbs.; Brown hair, brown eyes
EDUCATION: Bachelor of Science degree in mechanical engineering from Purdue University
MARITAL STATUS: Married to the former Betty L. Moore of Mitchell, Indiana.
CHILDREN: Scott, May 16, 1950; Mark, December 30, 1953
EXPERIENCE: Grissom is a lieutenant colonel in the United States Air Force, and received his wings in March 1951. He flew 100 combat missions in Korea in F-86's with the 334th Fighter-Interceptor Squadron. He left Korea in June 1952 and became a jet instructor at Bryan, Texas.

In August 1955, he entered the Air Force Institute of Technology at Wright-Patterson Air Force Base, Ohio, to study aeronautical engineering. In October 1956, he attended the Test Pilot School at Edwards Air Force Base, California, and returned to Wright-Patterson Air Force Base in 1957 as a test pilot assigned to the fighter branch.

Grissom has logged more than 4,000 hours flying time, including more than 3,000 hours in jet aircraft. He was awarded the Distinguished Flying Cross and the Air Medal with Cluster for service in Korea.

Grissom was named in April 1959 as one of the seven Mercury astronauts. He was the pilot of the Mercury-Redstone 4 (Liberty Bell 7) suborbital mission, July 21, 1961 and the command pilot of the Gemini 3 mission, March 23, 1965.

He is responsible for the Gemini group in the Astronaut Office, one of three organizational units in that office. (The others – Apollo and Operations).

Grissom is the son of Mr. and Mrs. Dennis Grissom, Mitchell, Indiana.

JOHN W. (for Watts) YOUNG, Gemini 6 backup pilot

BORN: San Francisco, California, September 24, 1930
HEIGHT: 5 feet 9 inches; WEIGHT: 172 lbs.; Brown hair, green eyes
EDUCATION: Bachelor of Science degree in aeronautical engineering from Georgia Institute of Technology
MARITAL STATUS: Married to the former Barbara V. White of Savannah, Georgia.
CHILDREN: Sandy, April 30, 1957; John, January 17, 1959
EXPERIENCE: Upon graduation from Georgia Tech, Young entered the United States Navy and is now a Commander in that service. From 1959 to 1962 he served as a test pilot, and later program manager of F4H weapons systems project, doing test and evaluation flights and writing technical reports.

He served as maintenance officer for all-weather Fighter Squadron 143 at the Naval Air Station, Miramar, California. In 1962, Young set world time-to-climb records in the 3,000 meter and 25,000 meter events in the F4B Navy fighter.

He has logged more than 3,200 hours flying time, including more than 2,700 hours in jet aircraft.

Young was among the group of nine astronauts selected by NASA in September 1962.
He was the pilot of Gemini 3, March 23, 1965.
He is the son of Mr. and Mrs. William H. Young, Orlando, Florida.

PREVIOUS GEMINI FLIGHTS

Gemini 1, April 8, 1964

Unmanned orbital flight, using first production spacecraft, to test Gemini launch vehicle performance and ability of launch vehicle and spacecraft to withstand launch environment. Spacecraft and second stage launch vehicle orbited for about four days. No recovery attempted.

Gemini 2, January 19, 1965

Unmanned ballistic flight to qualify spacecraft reentry heat protection and spacecraft systems. Delayed three times by adverse weather, including hurricanes Cleo and Dora. December launch attempt terminated after malfunction detection system shut engines down because of hydraulic component failure. Spacecraft recovered after ballistic reentry into Atlantic Ocean.

Gemini 3, March 23, 1965

First manned flight, with Astronauts Virgil I. Grissom and John W. Young as crew. Orbited Earth three times in four hours, 53 minutes. Landed about 50 miles (81 kilometers) short of planned landing area in Atlantic because spacecraft did not provide expected lift during reentry. First manned spacecraft to maneuver out of plane, after its own orbit. Grissom, who made suborbital Mercury flight, is first man to fly into space twice.

Gemini 4, June 3-7, 1965

Second manned Gemini flight completed 62 revolutions and landed in primary Atlantic recovery area after 97 hours, 59 minutes of flight. Astronaut James A. McDivitt was command pilot. Astronaut Edward H. White II was pilot, accomplished 21 minutes of Extravehicular Activity (EVA), using a hand held maneuvering unit for first time in space. Attempt to perform near-rendezvous with GLV second stage failed because of insufficient quantity of maneuvering fuel. Malfunction in Inertial Guidance System required crew to perform zero-lift reentry.

Gemini 5, August 21-29, 1965

Longest space flight on record. Astronauts L. Gordon Cooper and Charles (Pete) Conrad, Jr., circled the Earth 120 times in seven days, 22 hours and 59 minutes. Cooper is first to make two orbital space flights; has more time in space than any other human. Conrad, on first space flight, becomes world's second most experienced astronaut. Failure of oxygen heating system in fuel cell supply system threatened mission during first day of flight, but careful use of electrical power, and excellent operational management of fuel cells by both crew and ground personnel, permitted crew to complete flight successfully. Spacecraft landed about 100 miles (161 kilometers) from primary Atlantic recovery vessel because of erroneous base-line information programmed into onboard computer, although computer itself performed as planned. Plan to rendezvous with a transponder-bearing pod carried aloft by Gemini 5 was canceled because of problem with fuel cell oxygen supply.

Note: Gemini 6 previously was scheduled for launch October 25, 1965. The launch attempt was canceled when the Agena, with which Gemini 6 was to rendezvous, failed to achieve orbit despite a successful Atlas booster launch.

PROJECT OFFICIALS

George E. Mueller	Associate Administrator, Office of Manned Space Flight, NASA Headquarters, Acting Director, Gemini Program
LeRoy E. Day	Acting Deputy Director, Gemini Program, NASA Headquarters
William C. Schneider	Deputy Director, Mission Operations, Office of Manned Space Flight, NASA Headquarters, Gemini 7 and 6 Mission Director
Charles W. Mathews	Gemini Program Manager, Manned Spacecraft Center, Houston, Texas
Christopher C. Kraft	Flight Director, Manned Spacecraft Center, Houston
G. Merritt Preston	Deputy Mission Director for Launch Operations, John F. Kennedy Space Center, NASA., Kennedy Space Center, Florida
Lt. Gen. Leighton I. Davis	USAF, National Range Division Commander and DOD Manager of Manned Space Flight Support Operations
Maj. Gen. V. G. Huston	USAF, Deputy DOD Manager and Commander AFETR
Col. Richard C. Dineen	Director, Directorate Gemini Launch Vehicles, Space Systems Division, Air Force Systems Command
Lt. Col. John G. Albert	Chief, Gemini Launch Division, 6555th Aerospace Test Wing, Air Force Missile Test Center, Cape Kennedy, Florida
Rear Adm. W. C. Abhau	USN, Manned Spacecraft Support, Commander, Task Force 140 (Atlantic Ocean Recovery Forces)

U.S. MANNED SPACE FLIGHTS

Mission	Spacecraft Hrs.			Manned Hours In Mission			Total Manned Hrs. Cumulative		
	Hrs.	Min.	Sec.	Hrs.	Min.	Sec.	Hrs.	Min.	Sec.
MR-3 (Shepard)		15	22		15	22		15	22
MR-4 (Grissom)		15	37		15	37		30	59
MA-6 (Glenn)	4	55	23	4	55	23	5	26	22
MA-7 (Carpenter)	4	56	05	4	56	05	10	22	27
MA-8 (Schirra)	9	13	11	9	13	11	19	35	38
MA-9 (Cooper)	34	19	49	34	19	49	53	55	27
Gemini 3 (Grissom & Young)	4	53	00	9	46	00	63	41	27
Gemini 4 (McDivitt & White)	97	56	11	195	52	22	259	33	49
Gemini 5 (Cooper & Conrad)	190	56	01	381	52	02	641	25	51

SPACECRAFT CONTRACTORS

McDonnell Aircraft Corp., St. Louis, Mo., is prime contractor for the Gemini spacecraft. Others include:

AIResearch Manufacturing Co., Los Angeles, Calif. . . . . Environment Control System
The Eagle Pitcher Co., Joplin, Mo. . . . . Batteries
IBM Corp., New York, N.Y. . . . . Computer, Guidance
Northrop Corp., Newbury Park, Calif. . . . . Parachutes
Rocketdyne, Canoga Park, Calif. . . . . OAMS, RCS
Thiokol Chemical Corp., Elkton, Md. . . . . Retrorocket System
Weber Aircraft Corp., Burbank, Calif. . . . . Ejection Seats
Westinghouse Electric Corp., Baltimore, Md. . . . . Rendezvous Radar System

Atlas contractors include:

General Dynamics, Convair Div., San Diego, Calif.	Airframe and Systems Integration
Rocketdyne Div., North American Aviation, Inc., Canoga Park, Calif.	Propulsion Systems
General Electric Co., Syracuse, New York	Guidance

Titan II contractors include:

Martin Co., Baltimore Divisions, Baltimore, Md.	Airframe and Systems Integration
Aerojet-General Corp., Sacramento, Calif.	Propulsion Systems
General Electric Co., Syracuse, N.Y.	Radio Command Guidance System
Burroughs Corp., Paoli, Pa.	Ground Guidance Computer
Aerospace Corp., El Segundo, Calif.	Systems Engineering and Technical Direction

Agena D contractors include:

Lockheed Missiles and Space Co., Sunnyvale, Calif.	Airframe and Systems Integration
Bell Aerosystems Co., Niagara Falls, N.Y.	Propulsion Systems
McDonnell Aircraft Co., St. Louis, Mo.	Target Docking Adapter

Food contractors:

U.S. Army Laboratories, Natick, Mass.	Food Formulation Concept
Whirlpool Corp., St. Joseph, Mich.	Procurement, Processing, Packaging
Swift and Co., Chicago and Pillsbury Co., Minneapolis	Principal Food Contractors

Suit contracts:
The David R. Clark Co., Worcester, Mass.

GODDARD SPACE FLIGHT CENTER GREENBELT, MARYLAND

GT 7-6 WORLD-WIDE COMMUNICATIONS NETWORK

TRACKING AND DATA ACQUISITION OF TWO SPACECRAFT

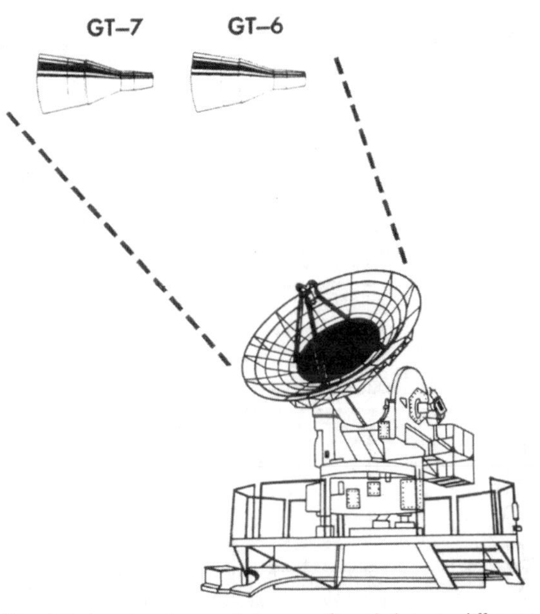

C-Band Radar pinpoints each spacecraft and detects difference from variations in beacon signals.

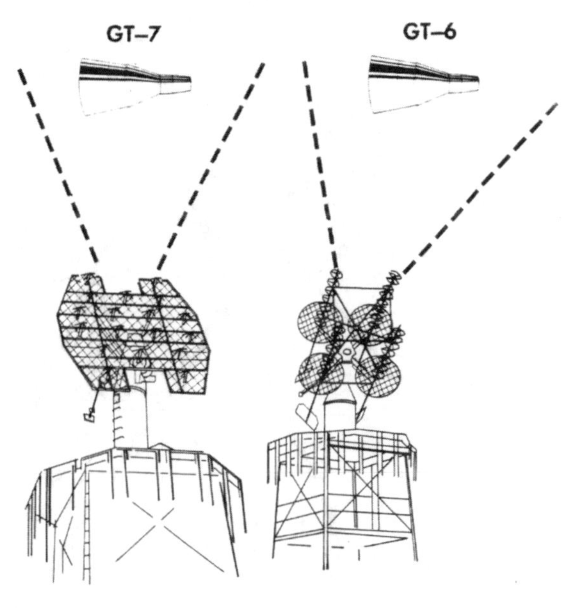

All radio signals bearing information about men and capsule enter the PCM telemetry antenna. GT-7 data is directed to Gemini Systems Monitor Console; GT-6 data to Agena Systems Monitor Console.

AGENA SYSTEMS CONSOLE
GT-6 spacecraft data will be displayed on Agena Systems Monitor Console.

COMMAND COMMUNICATOR CONSOLE
Presents composite status of spacecraft systems and has ground command controls for triggering on-board systems events.

GEMINI SYSTEMS CONSOLE
GT-7 spacecraft data is displayed on Gemini Systems Monitor Console throughout mission.

AEROMEDICAL MONITOR CONSOLE. Full biomedical presentations will be displayed at key ground station Aeromedical Consoles during the GT-7 portion of the mission. For the GT7-6 rendezvous segment, composite data on the four astronauts will be presented at the console with the remainder going to Sanborn recorders and Cardiotac (meters) counters located next to the console.

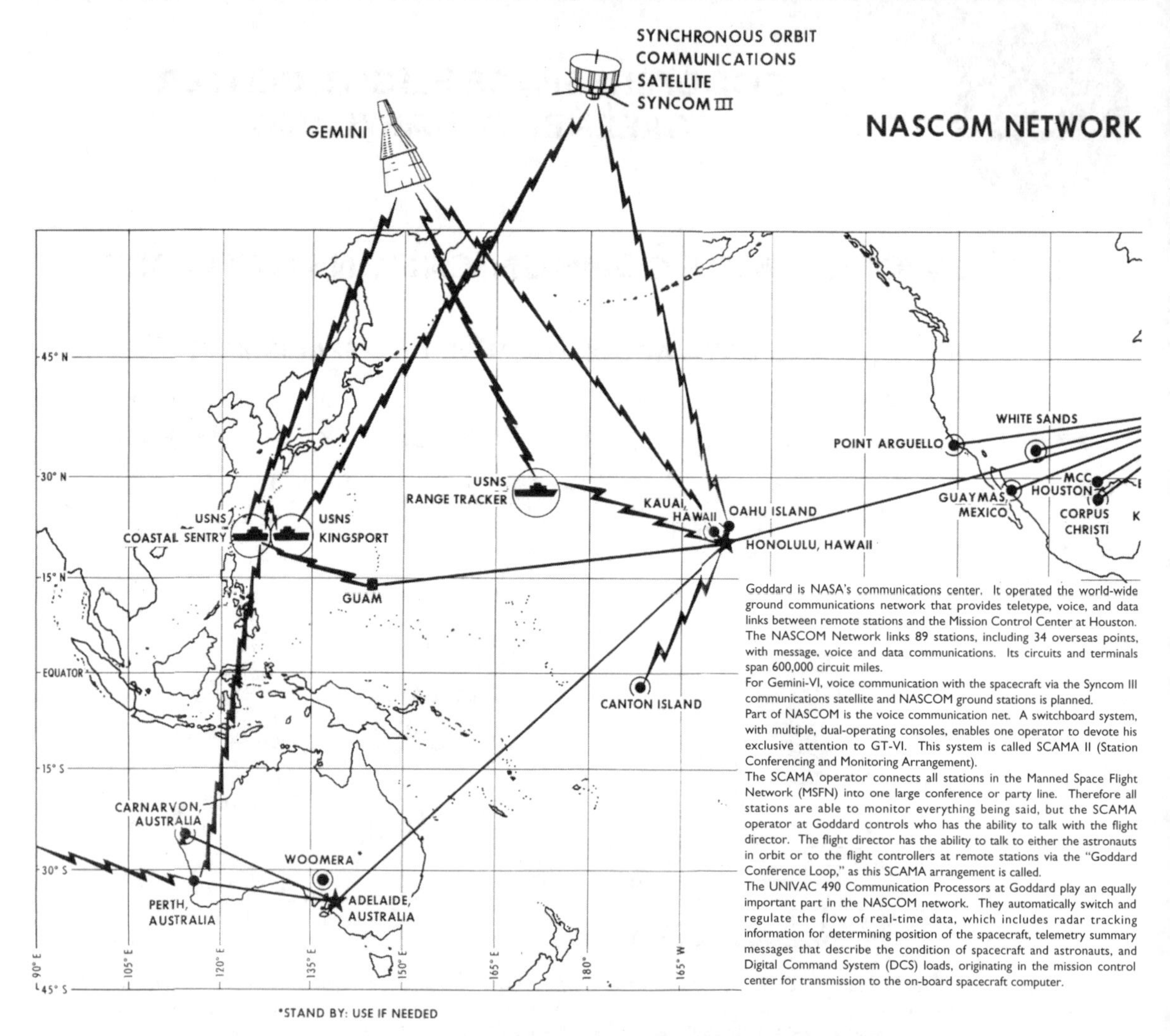

Goddard is NASA's communications center. It operated the world-wide ground communications network that provides teletype, voice, and data links between remote stations and the Mission Control Center at Houston. The NASCOM Network links 89 stations, including 34 overseas points, with message, voice and data communications. Its circuits and terminals span 600,000 circuit miles.

For Gemini-VI, voice communication with the spacecraft via the Syncom III communications satellite and NASCOM ground stations is planned.

Part of NASCOM is the voice communication net. A switchboard system, with multiple, dual-operating consoles, enables one operator to devote his exclusive attention to GT-VI. This system is called SCAMA II (Station Conferencing and Monitoring Arrangement).

The SCAMA operator connects all stations in the Manned Space Flight Network (MSFN) into one large conference or party line. Therefore all stations are able to monitor everything being said, but the SCAMA operator at Goddard controls who has the ability to talk with the flight director. The flight director has the ability to talk to either the astronauts in orbit or to the flight controllers at remote stations via the "Goddard Conference Loop," as this SCAMA arrangement is called.

The UNIVAC 490 Communication Processors at Goddard play an equally important part in the NASCOM network. They automatically switch and regulate the flow of real-time data, which includes radar tracking information for determining position of the spacecraft, telemetry summary messages that describe the condition of spacecraft and astronauts, and Digital Command System (DCS) loads, originating in the mission control center for transmission to the on-board spacecraft computer.

*STAND BY: USE IF NEEDED

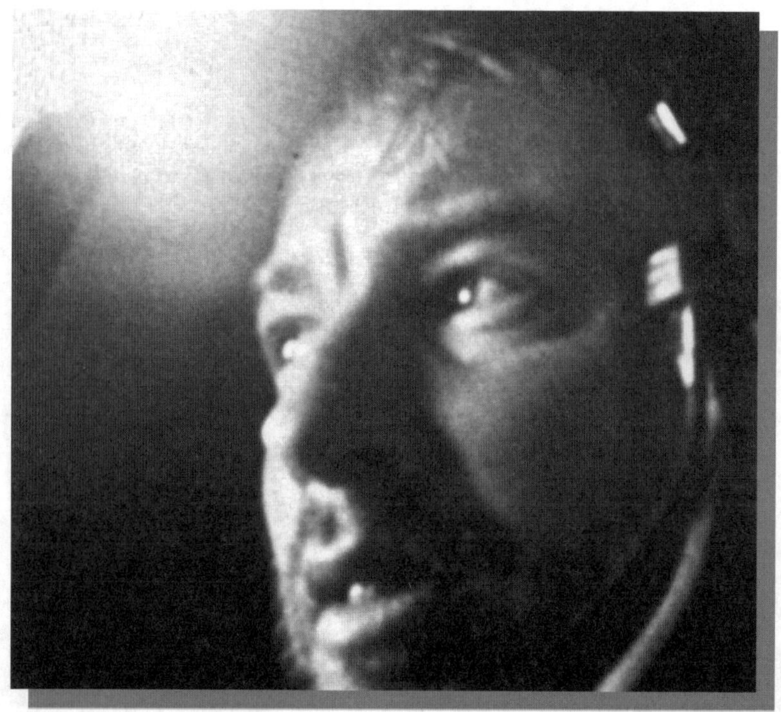

SUPPORT FOR GT 7-6

★ COMMUNICATIONS SWITCHING CENTER
○ MANNED SPACE FLIGHT TRACKING STATION
——— CABLE
■ COMMUNICATIONS TERMINAL
RADIO

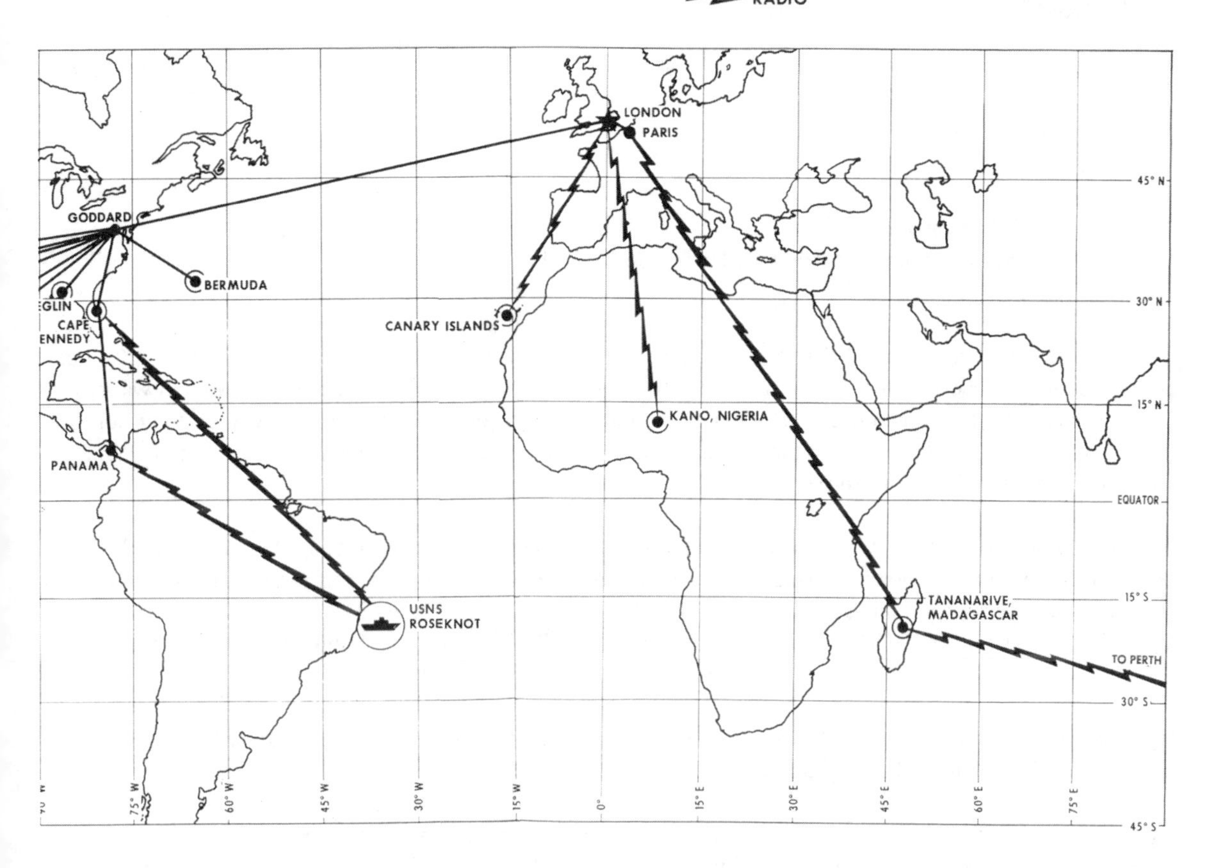

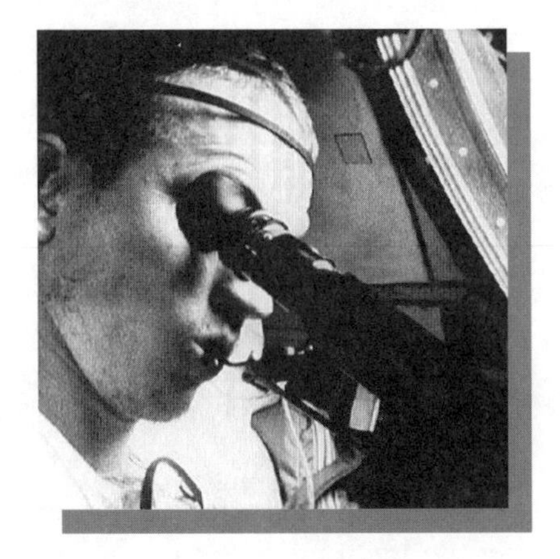

NATIONAL AERONAUTICS AND SPACE ADMINISTRATION
WASHINGTON, D.C. 20546

Mission Operation Report
No. M-913-65-07

MEMORANDUM

November 18, 1965

To: A / Administrator
From: M / Associate Administrator for Manned Space Flight
Subject: Gemini VII Mission

Gemini flight mission number VII is scheduled to be launched from Complex 19 at the John F. Kennedy Space Center on December 4, 1965. It will be the fourth manned flight in the Gemini series and will have a mission duration of fourteen days. The primary operational objective of Gemini VII will be long duration flight. In addition, it will serve as a target vehicle for Gemini mission VI-A which is planned to be launched nine days after the Gemini VII mission.

Nominal launch time of the Gemini launch vehicle (GLV) will be at 2:30 p.m. EST. Launch azimuth of the GLV will be 83.6 degrees with a resultant orbital inclination of 28.87 degrees, and perigee-apogee of 87-183 nautical miles. A perigee adjustment during the third revolution will provide an orbit lifetime of at least fifteen days. The flight will be approximately 13 days, 17 hours and 30 minutes in duration with 206 revolutions completed, and recovery occurring at the beginning of the 207th revolution.

Frank Borman and James A. Lovell will serve as primary crew for this mission. Backup will be provided by Edward H. White II and Michael Collins.

Twenty experiments will be performed during the flight in addition to the rendezvous activity associated with Gemini mission VI-A. Spacecraft electrical power will be provided by fuel cells as was Gemini V. Spacecraft recovery is expected to occur 360 nautical miles south of Bermuda in the West Atlantic zone at 8:00 a.m. on December 18. Recovery operations, as with Gemini VI-A will be televised in real time by commercial television from the aircraft carrier. As in previous missions, post-flight debriefings will be conducted aboard the carrier, at Kennedy Space Center and at the Manned Spacecraft Center.

George E. Mueller

Report No. M-913-65-07

MISSION OPERATION REPORT

GEMINI VII FLIGHT

OFFICE OF MANNED SPACE FLIGHT

FOREWORD

MISSION OPERATION REPORTS are published expressly for the use of NASA General Management as required by the Administrator in NASA Instruction 6-2-10 dated August 15, 1963. The purpose of these reports is to provide NASA General Management with timely, complete and definitive information on flight mission plans and results from launchings with Scout class or larger vehicles.

Initial reports are to be prepared and issued for each flight project just prior to launch. Following launch, updating reports for each mission will be issued to keep General Management currently informed as provided in NASA Instruction 6-2-10.

Distribution of these reports has been specifically directed by General Management and they are not available for additional or general distribution. The Office of Public Affairs publishes a comprehensive series of pre-launch and post-launch reports on NASA flight missions which are available for general distribution.

Published and Distributed
by
OFFICE OF PROGRAM REPORTS
OFFICE OF PROGRAMMING
NATIONAL AERONAUTICS AND SPACE ADMINISTRATION
Washington, D.C. 20546

GENERAL

The Gemini Program is a continuation of the study of man's adaptation to the space environment and the development of operational procedures for future space flights. This includes studying the crew while under conditions of zero gravity for extended periods, making simple repairs and performing manual exercises, maneuvering the spacecraft for rendezvous and docking with other orbiting vehicles, and controlling the spacecraft during reentry.

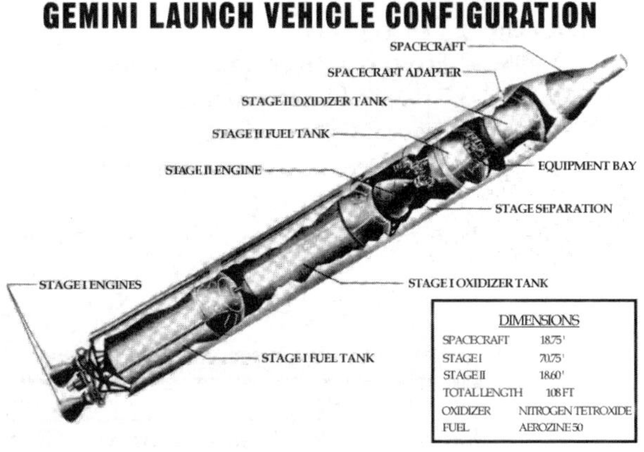

FIG. I

Gemini flight number seven will be the fourth manned orbital flight in the Gemini Program. Previous missions have proven the success of orbital insertion, structural integrity, and spacecraft systems performance. Gemini IV demonstrated the feasibility of a four-day mission duration and extravehicular activity. On Gemini V an eight-day duration was demonstrated and the radar and guidance system was further evaluated. Gemini VII will be a long duration mission and will have a flight time of 13 days, 17 hours and 30 minutes. Figure 1 illustrates the Gemini launch vehicle and spacecraft.

GEMINI VII MISSION OBJECTIVES

PRIMARY

- Demonstrate manned orbital flight in the Gemini spacecraft for a period of approximately fourteen days.
- Evaluate the effects of exposing the crew to longer periods of weightlessness.

SECONDARY

- Utilize the Gemini 7 spacecraft as a rendezvous target for the Gemini VI-A mission.
- Maneuver the spacecraft to station keep with the GLV second stage from SECO+30 seconds until approximately five minutes after entering the first darkness period (approximately twenty-five minutes).
- Demonstrate a controlled reentry to a predetermined landing point.
- Evaluate the lightweight suit (G-5C).
- Execute or perform the following assigned experiments:

  Medical Experiments
  1. M-1 Cardiovascular Conditioning
  2. M-3 In-flight Exerciser
  3. M-4 In-flight Phonocardiogram
  4. M-5 Bioassays of Body Fluids
  5. M-6 Bone Demineralization
  6. M-7 Calcium Balance Study
  7. M-8 In-flight Sleep Analysis
  8. M-9 Human Otolith Function

  Technological Experiments
  1. D-4 Celestial Radiometry
  2. D-5 Star Occultation
  3. D-7 Radiometric Observation of Objects in Space
  4. D-9 Simple Navigation
  5. D-13 Flight Crew Visibility

Scientific Experiments
1. S-5 Synoptic Terrain Photography
2. S-6 Synoptic Weather Photography
3. S-8 Visual Acuity

Engineering Experiments
1. MSC-2 Proton-Electron Spectrometer
2. MSC-3 Tri-Axis Magnetometer
3. MSC-4 Optical Communications
4. MSC-12 Landmark Contrast Measurements

Objectives of later missions in the Gemini Program are summarized in the following table:

TABLE I

Mission	Objectives	Date
Gemini VI-A	Rendezvous with Gemini VII Spacecraft	Dec '65
Gemini VIII	Mode II Rendezvous Development	2Q66
Gemini IX	Simultaneous Countdown and Rendezvous	3Q66
Gemini X	LEM Rendezvous Simulation	3Q66
Gemini XI	Rendezvous Advancement	4Q66
Gemini XII	LEM Rendezvous Simulation	1Q67

DESCRIPTION OF GEMINI LAUNCH VEHICLE

The Gemini launch vehicle has been modified by man-rating a Titan II missile. It consists of two stages which are ten feet in diameter and have a total length of 90 feet. Gross weight is approximately 345,000 pounds. The first and second stage engines produce thrusts of 430,000 and 100,000 pounds, respectively. The Gemini launch vehicle for the Gemini VII mission is of the same configuration as used for Gemini V except for the four flashing lights, similar to the ones installed on GLV IV, which will be installed on GLV VII to aid in the station-keeping exercise. The systems have been discussed in previous reports, therefore a description will not be included in this report.

DESCRIPTION OF GEMINI SPACECRAFT

The spacecraft as shown in Figure 2 to be used for the Gemini VII mission will be configured for a 14-day long-duration mission. The configuration will be much the same as that of Gemini V with the exception of the rendezvous evaluation pod (REP) and associated REP equipment which was on spacecraft 5. These will not be installed on spacecraft 7. Fuel cells similar to the ones used on spacecraft 5 will be used instead of the batteries used on spacecraft 6. In addition an auxiliary fuel tank will be installed to carry additional OAMS fuel. Two flashing lights will be installed on the spacecraft adapter section, and an L-band transponder will be installed in the Gemini 7 rendezvous and recovery section for use as rendezvous aids by the Gemini 6 crew.

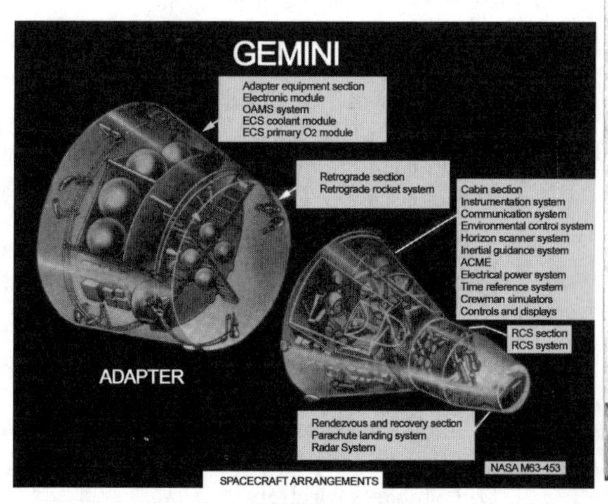

FIG. 2

TABLE II

CONSUMABLES LOADING		
	LOADED	USABLE
Oxygen		
Primary	106 lbs.	
Reentry Module	14 lbs.	
OAMS		
Fuel	177.1 lbs.	172.2lbs.
Oxidizer	245.9 lbs.	234.5 lbs.
RCS		
Fuel	31.6 lbs.	27.6 lbs.
Oxidizer	40.4 lbs.	36.0 lbs.
Food	36.4 lbs. (28 man-day packs)	
Drinking Water		
Adapter	174.0 lbs.	
Reentry Module	14.6 lbs.	
Fuel Cell Reactants		
Oxygen	183.6 lbs.	
Hydrogen	23.6 lbs.	

EXPERIMENTS

The twenty experiments scheduled for Gemini VII are described below. Fourteen of these experiments were performed on previous missions and six will be flown for the first time on this flight.

Experiments Flown on Earlier Missions

1. M-I, Cardiovascular Conditioning

This experiment will determine the effectiveness of cyclic inflation of pneumatic venous pressure on the thighs as a prevention measure of cardiovascular deconditioning induced by prolonged weightlessness. Pneumatic pressure cuffs, integral to the spacesuit, are located around the thighs and inflated periodically to 80mm of mercury pressure increasing venous pressure below the cuffs. The automatic pressurization cycle lasts two minutes out of every six and will utilize oxygen from the environmental control system. Figure 3 shows the pneumatic cuff in a test configuration employing oxygen from a pressure bottle.

M-1 CARDIOVASCULAR CONDITIONING (PNEUMATIC CUFFS)

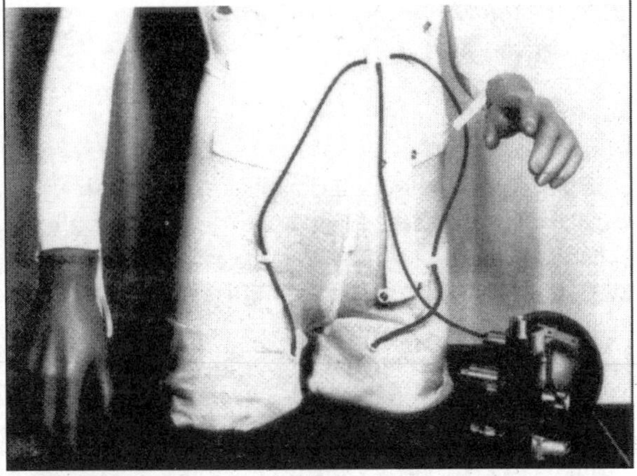

FIG. 3

2. M-3, In Flight Exerciser

The objective of this experiment is to assess the astronauts' capacity to perform physical work and their capability for sustained performance. The rapidity with which the heart rate returns to normal after cessation of exercise is an indication of an individual's physical fitness. A workload will be provided by specific periods of exercise at the rate of one pull per second for 30 seconds on an exercise device that requires a known amount of effort.

The exercise device consists of a pair of bungee cords attached to a nylon handle at one end and a nylon foot strap at the other end. Figure 4 shows the exercise device and the manner in which this exercise will be performed.

The in-flight data obtained will be compared with the control data e.g., exercises performed prior to the mission under monitored conditions to determine the capacity for work in space.

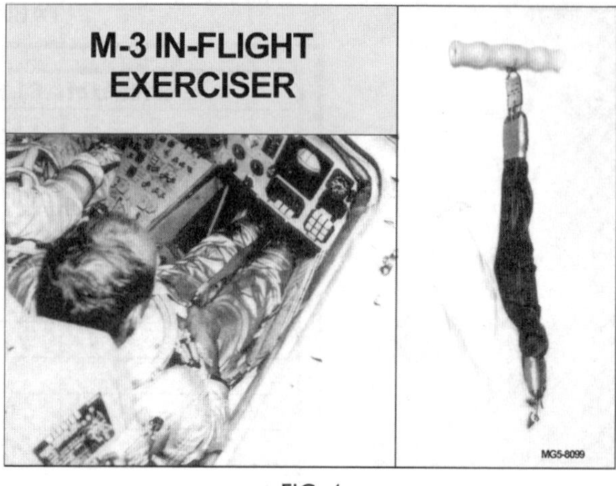

FIG. 4

3. M-4, In-Flight Phonocardiogram

In this experiment the fatigue state of an astronaut's heart muscle will be determined by measuring the time interval between the electric activation of a muscle and the onset of muscular contraction.

A microphone will be applied to an astronaut's chest wall at the cardiac apex. Heart sounds detected during the flight will be recorded on an onboard biomedical recorder. The sound trace will be compared to the waveform obtained from a simultaneous inflight electrocardiogram to determine the time interval between electrical activation of the heart muscle and the onset of ventricular systole. Figure 5 illustrates the method of installation of the phonocardiogram transducer.

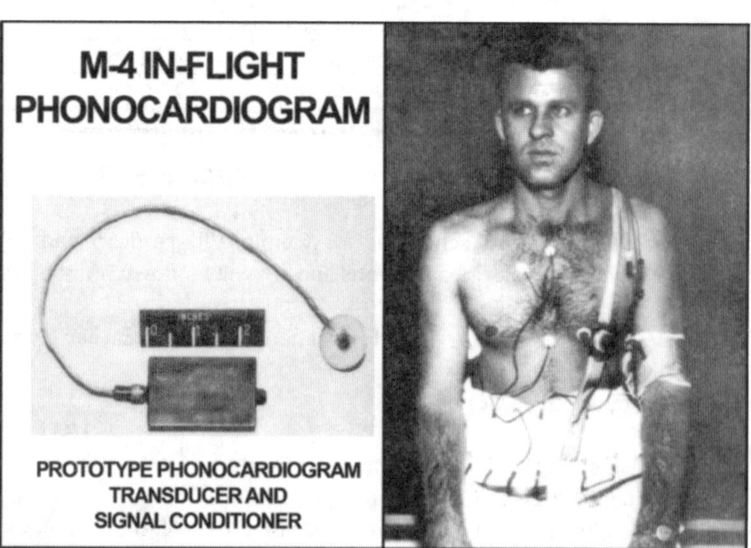

FIG. 5

4. M-6, Bone Demineralization

The purpose of this experiment is to establish the occurrence and degree of bone demineralization influenced by the relative immobilization associated with the cockpit of the Gemini spacecraft and weightlessness.

Special X-rays will be taken of an astronaut's heel bone and the terminal bone of the fifth digit of the right hand. Three pre-flight and three post-flight exposures will be taken of these two bones and compared to determine if any bone demineralization has occurred due to the space flight.

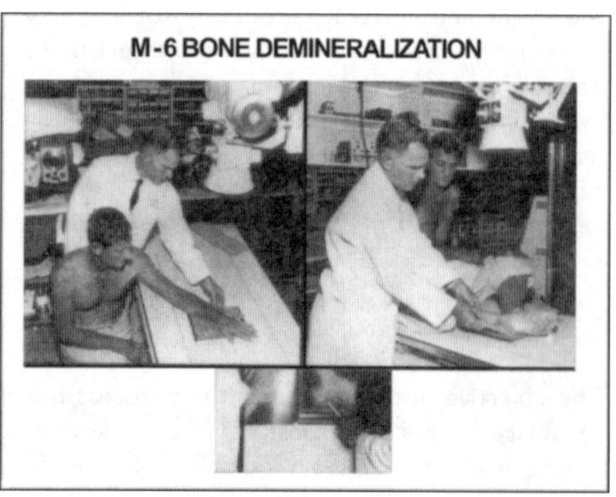

FIG. 6

The equipment to be utilized in this experiment will be closely calibrated clinical X-ray machines, standard 11-inch by 14-inch X-ray films and calibrated wedge densitometers. Figure 6 illustrates the laboratory equipment that will be used.

5. M-9, Human Otolith Function

The purpose of this experiment is to measure changes and determine orientation capability in the otolith (gravity sensors in the inner ear) function during prolonged weightlessness. When certain procedures are followed, the conjugate rolling of the eyes around their lines of rotation opposite to the lateral inclination of the head (ocular counterrolling) is generally held to be a direct reflex originating in the otolith organs. Ocular counterrolling will be measured on the astronauts before and after the mission period. Measurements will also be made on the egocentric visual localization (EVL) of a spacecraft frame of reference before, during and after flight as an indicator of orientation within various gravitoinertial environments. Figure 7 illustrates the special type goggles and tilt table used in conjunction with this experiment.

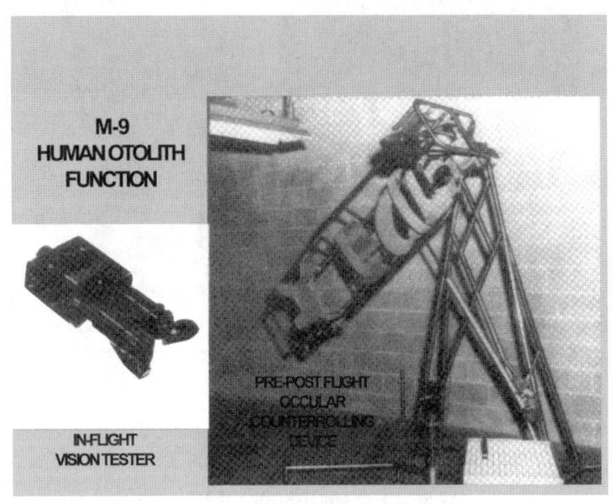

FIG. 7

6. MSC-2, Proton-Electron Spectrometer

In order to determine the degree of hazard, if any, to which the crew will be subjected on space flights, it is necessary to project what radiation environment any given mission will encounter. Specifically, this experiment will make measurements external to the spacecraft in a region where the inner Van Allen radiation belt dips close to the Earth's surface due to the irregular strength of the Earth's magnetic field, and is usually referred to as the South Atlantic Geomagnetic Anomaly.

FIG. 8

This measurement will be accomplished by means of a scintillating-crystal, charged-particle analyzer mounted on the adapter assembly of the spacecraft. Data from this experiment will be used to correlate radiation measurements made inside the spacecraft and to predict radiation levels on future space missions. The proton electron spectrometer installation is shown in Figure 8.

7. MSC-3, Tri-Axis Magnetometer

The purpose of this experiment is to monitor the direction and amplitude of the Earth's magnetic field with respect to an orbiting spacecraft. The astronauts will operate an adapter-mounted tri-axis fluxgate magnetometer as it passes through the South Atlantic Geomagnetic Anomaly. The magnitude of the three orthogonal components of the Earth's magnetic field will be measured with respect to the spacecraft. The measurement will be performed in conjunction with experiment MSC-2, Proton Electron Spectrometer, to determine the field, line direction and pitch angle of the impacting particles. Figure 9 illustrates the equipment to be used on this experiment.

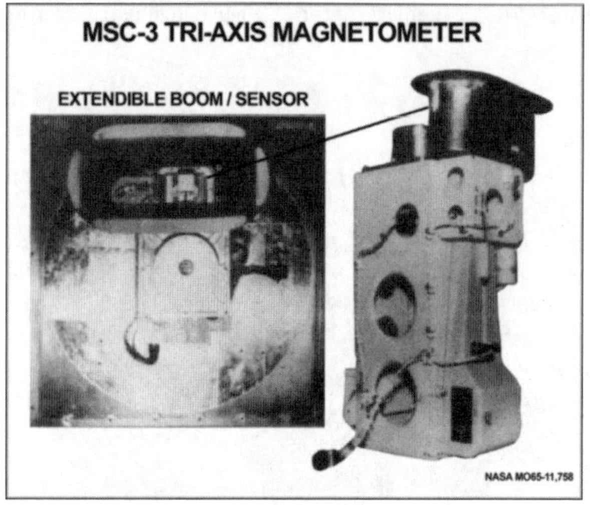

FIG. 9

8. D-4, Celestial Radiometry – D-7, Space Object Radiometry

The results of these experiments will provide information on the radiation intensity of celestial bodies and various objects in space. Instrumentation includes a three-channel spectra-radiometer, a dual-channel Michelson Interferometer-Spectrometer, and a cryogenically cooled Spectrometer. The equipment can measure radiant intensity from the ultra-violet through the infrared region. The sensing units are housed in the Gemini adapter section and are directed toward the objective by orienting the spacecraft. Figures 10 and 11 illustrate the location of equipment and the experiments in use.

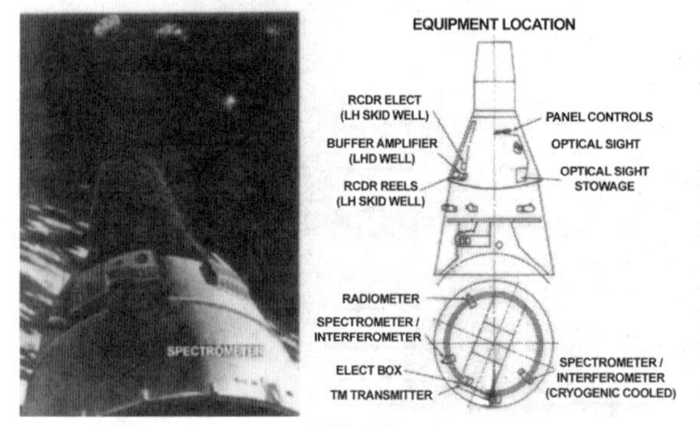

FIG. 10

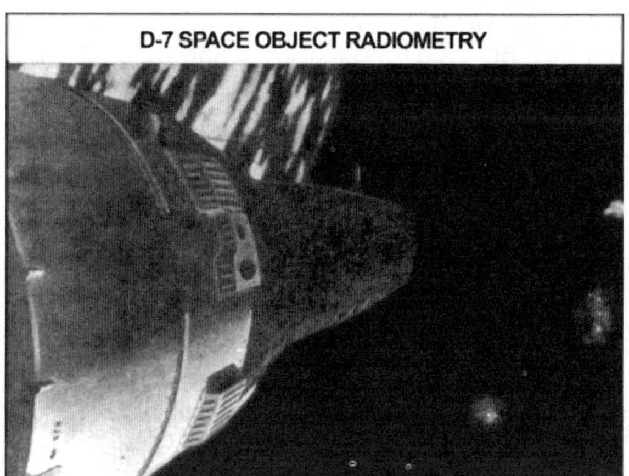

FIG. 11

9. D-9, Simple Navigation

The capability of man to navigate in space and to provide a reliable navigation system independent of ground support will be tested in this experiment. Two special instruments have been developed for use on Gemini spacecraft to allow detailed manual-visual examination of the space phenomena thought to be best for space navigation purposes. These are a space stadimeter and a sextant. This flight will only carry the space sextant which the astronaut will use to make star-horizon angular measurements for orbital orientation determinations. The results will be compared with actual parameters to determine the accuracy of the procedures. A hand-held sextant is shown in Figure 12.

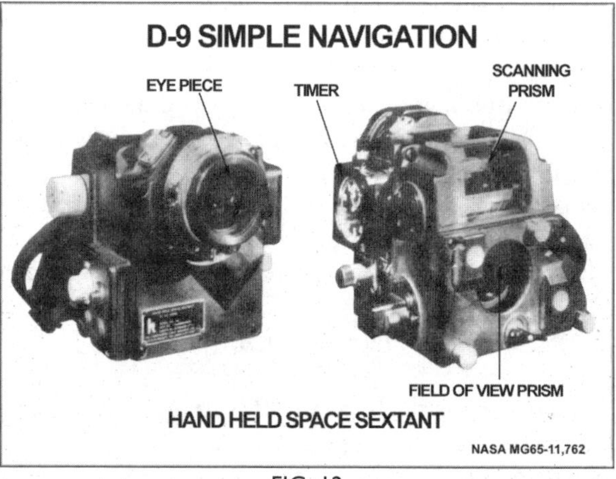

FIG. 12

10. S-5, Synoptic Terrain Photography

The objective of this experiment is to obtain high quality photographs of selected parts of the Earth's surface. The spacecraft will be manually oriented from an orbit mode attitude to a moderately high camera depression angle attitude. After a series of photographs has been taken, the spacecraft will be reoriented to the orbit mode attitude. Several of these spacecraft orientation maneuvers will be performed throughout the flight, during which approximately ninety pictures will be taken over areas of the world. Primary areas are the shallow waters around the Bahamas, the Red Sea, and the west central portion of Mexico. Figure 13 shows one such photograph taken on Gemini 4.

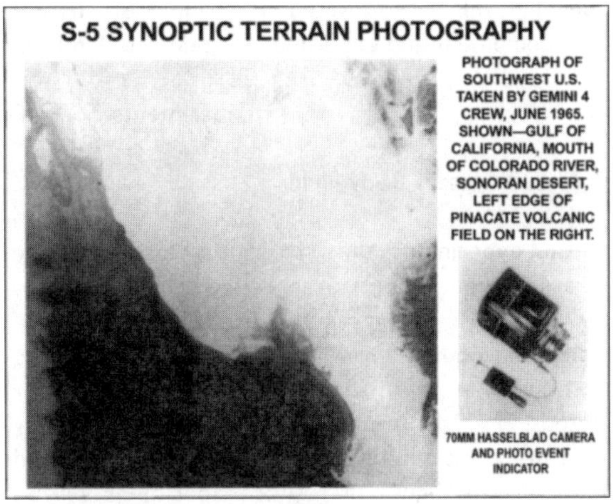

FIG. 13

11. S-6, Synoptic Weather Photography

The purpose of this experiment is to learn more about the Earth's weather system by obtaining high quality photographs of selected cloud formations. As in experiment S-5, the spacecraft will be oriented from an orbit mode attitude to a moderately high camera depression angle attitude. After a series of photographs has been taken, the spacecraft will be reoriented to the orbit mode attitude. Several orientation maneuvers will be required during which approximately ninety pictures will be taken. The photograph shown in Figure 14 was taken on Gemini IV and is similar to those expected from this flight.

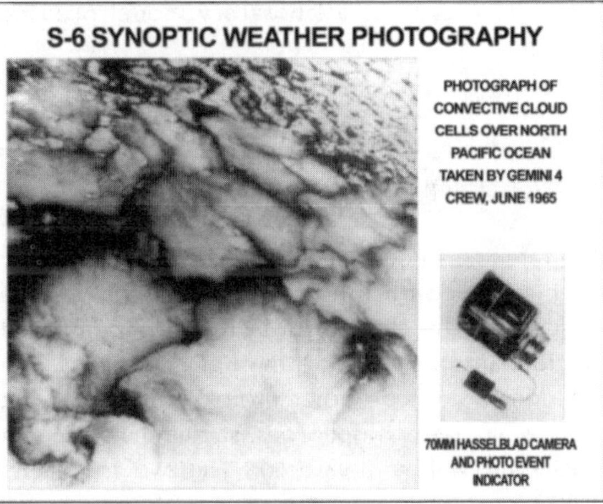

FIG. 14

12. S-8, Visual Acuity – D-13, Astronaut Visibility

The visual ability of the astronauts in the detection and recognition of objects on the Earth's surface will be tested in these experiments. The spacecraft will be equipped with a vision tester and a photometer. The astronauts will use the vision tester to evaluate visual functions in space relative to Earthbound baseline values. The photometer will measure light attenuation of the spacecraft window due to scattering. While

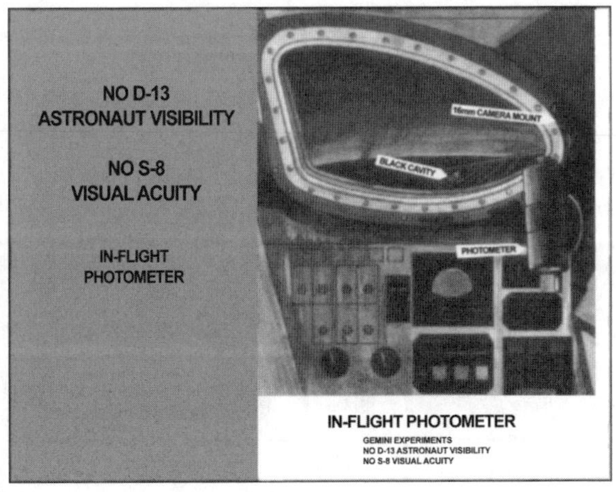

FIG. 15

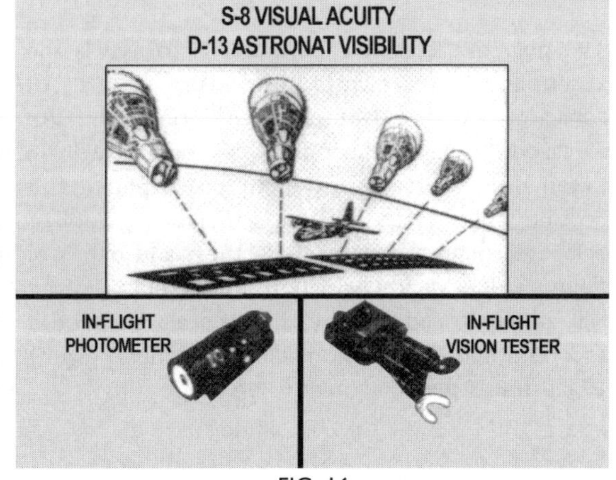

FIG. 16

the spacecraft is oriented the astronauts will view selected ground objects and record their findings. Viewings will be correlated with laboratory experiments and vision will be checked pre- and post-flight. The equipment and experiment are depicted in Figures 15 and 16.

Experiments To Be Flown For The First Time

1. M-5, Bioassays Body Fluids

In this experiment, the astronaut's reaction to the stress of requirements of space flight will be studied by means of analyzing the body fluids. Urine collection equipment will be utilized for obtaining in-flight specimens and for measuring urinary output. In-flight urine samples will be obtained by measuring the amount of urine at each voiding and reserving an aliquot of 100 cc. Specimens will be stored in special bags reserved for this purpose. Figure 17 illustrates a laboratory model of urine collection.

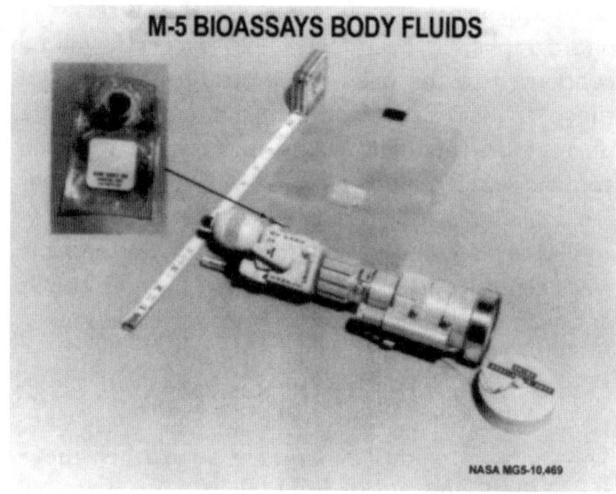

FIG. 17

2. M-7, Calcium Balance Study

The rate and amount of calcium change to the body during the conditions of orbital flight will be evaluated in this experiment by means of controlled calcium intake and output measurements. In addition to calcium, other electrolytes of interest such as nitrogen phosphorous, sodium chloride and magnesium will be monitored. The two astronauts will be maintained on a prescribed calcium diet for two weeks prior to, during and after flight. Careful recording of input-output will be accomplished and total fecal specimens will be preserved for analysis. Figure 18 shows equipment to be used to store output specimens.

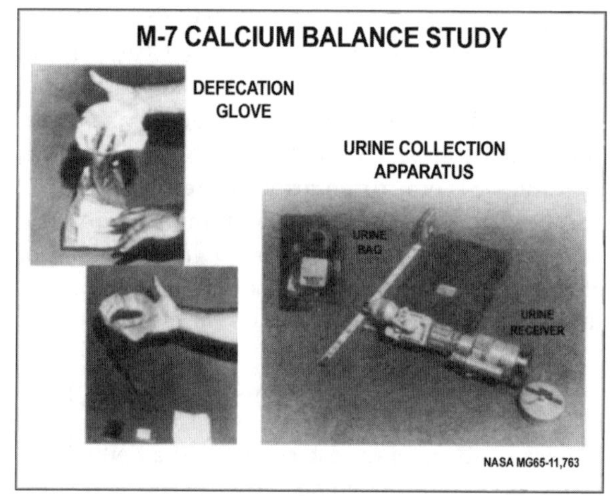

FIG. 18

3. M-8, In-Flight Sleep Analysis

The objective of this experiment is to assess the astronauts' state of alertness, levels of consciousness, and depth of sleep during flight. An electroencephalograph on the astronauts will be taken during weightless flight to establish the possible utility of the EEG as a monitoring tool to help determine the state of alertness and depth of sleep. The electrical activity of the cerebral cortex will be monitored by two pairs of scalp electrodes and recorded on the biomedical recorder. Figure 19 illustrates the equipment to be used.

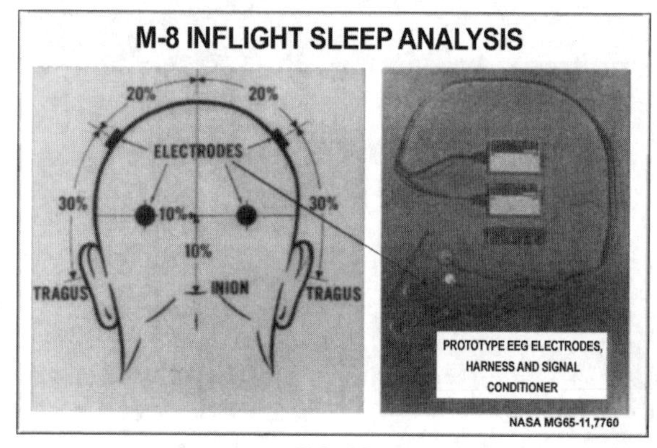

FIG. 19

4. MSC-4, Optical Communication

The purpose of this experiment is to establish optical communications link and demonstrate the use of optical frequencies for communications between an orbiting spacecraft and a pre-determined ground station. The pilot will maintain the spacecraft attitude while the co-pilot aims a hand-held laser transmitter at a visible light beam directed at the spacecraft from the ground. Scientific data on sky radiance and atmospheric transmission effects on optical frequencies will be recorded. An evaluation of man's efficiency as an element in a pointing servo loop will also be tested. Figure 20 illustrates the laser and ground based equipment.

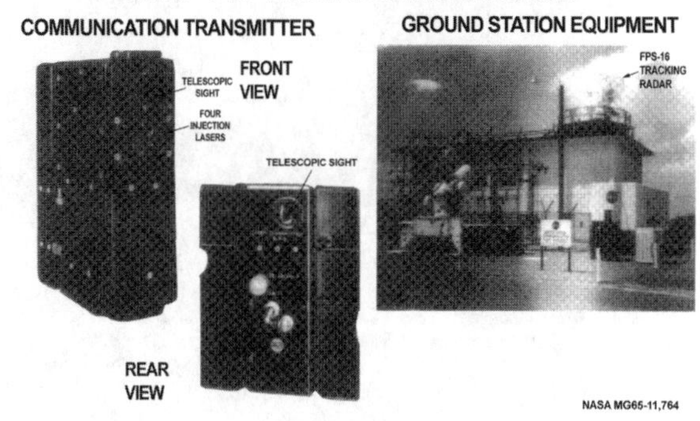

FIG. 20

5. MSC-12, Landmark Contrast Measurements

The purpose of this experiment is to measure the visual contrast of land-sea boundaries and other types of terrain to be used as a service of navigation data for the on-board Apollo Guidance and Navigation system. Landmark contrast measurements made from outside the atmosphere will provide data of a high confidence level to effectively duplicate navigation sightings for Apollo. Landmark measurements will be made of such areas as the Florida Coast, South American-Chilean Coast, African-Atlantic Coast, and Australian Coast. A helmet-mounted photometric telescope sensor and equipment used for the D-5 Star Occultation Navigation experiment will be used. Figure 21 depicts operation of this experiment equipment.

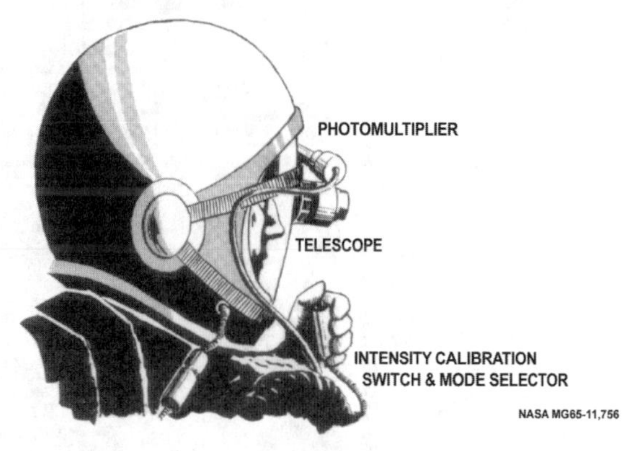

FIG. 21

6. D-5, Star Occultation Navigation

The feasibility and operational value of star occulting measurements in the development of a simple, accurate, and self-contained orbital navigational capability will be investigated by this experiment. The astronauts will determine the orbit of the Gemini spacecraft by measuring the time six stars dip behind an established horizon. The astronaut points the telescope at the star and centers it within a reticule circle. A portion of the radiated light is then diverted to a photomultiplier. A calibration mode is initiated by a hand-held switch and the star intensity is measured automatically. This information will be telemetered to the ground and the resulting orbit determination can be accomplished and compared with ground tracking data. Figure 22 illustrates the method of measuring the star occultation.

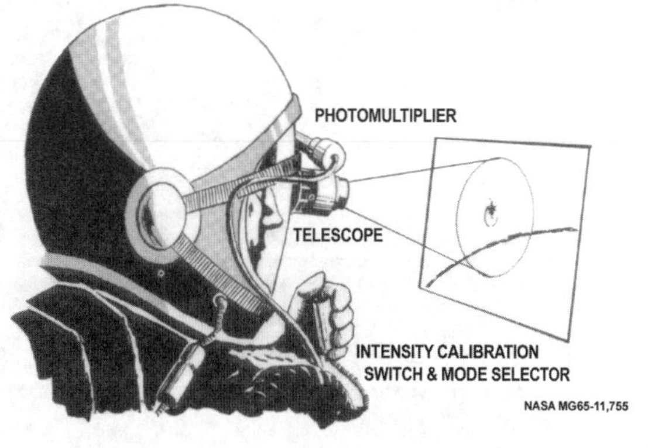

FIG. 22

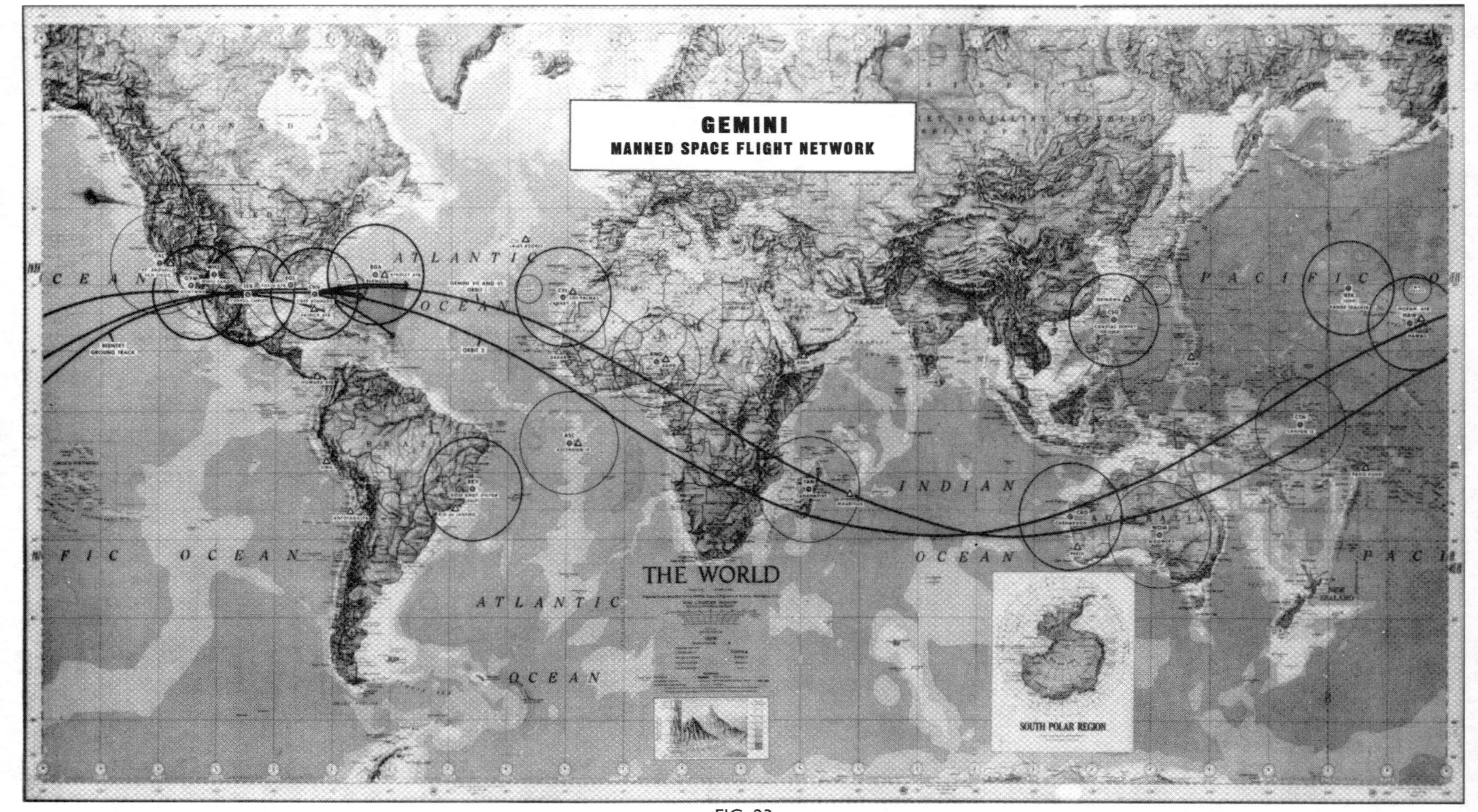
GEMINI
MANNED SPACE FLIGHT NETWORK
THE WORLD
SOUTH POLAR REGION
ATLANTIC OCEAN
INDIAN OCEAN

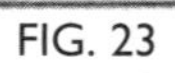
FIG. 23

In addition to the rendezvous and experiments, an operational exercise for the Apollo Landmark Investigation is scheduled if adequate OAMS fuel is available. The objective of this exercise is to evaluate selected Apollo landmarks for suitability as navigation references and to obtain photographs of landmarks for use as identification aids on Apollo missions.

GROUND OPERATIONAL SUPPORT SYSTEM

The Ground Operational Support System (GOSS) is the total complex of operational support systems on (or near) the Earth for tracking, obtaining data from, exercising control of, and communicating with manned space flight vehicles and consists of Launch Instrumentation, Mission Control Systems, and the Manned Space Flight Network (MSFN). The Air Force Eastern Test Range and certain of the MSFN stations provide elements of the Launch Instrumentation. Mission control will be effected from the Mission Control Center at Houston, (MCC-H). Figure 23 illustrates the MSFN station locations and the limits of orbital track which the network will cover. The communications interfaces between the MSFN Stations and the capabilities of each of these stations are reflected in Figure 24 and Table III, respectively. It is noted that capabilities exist which are not required for support of the subject mission (i.e. S-Band Radar and Agena Telemetry).

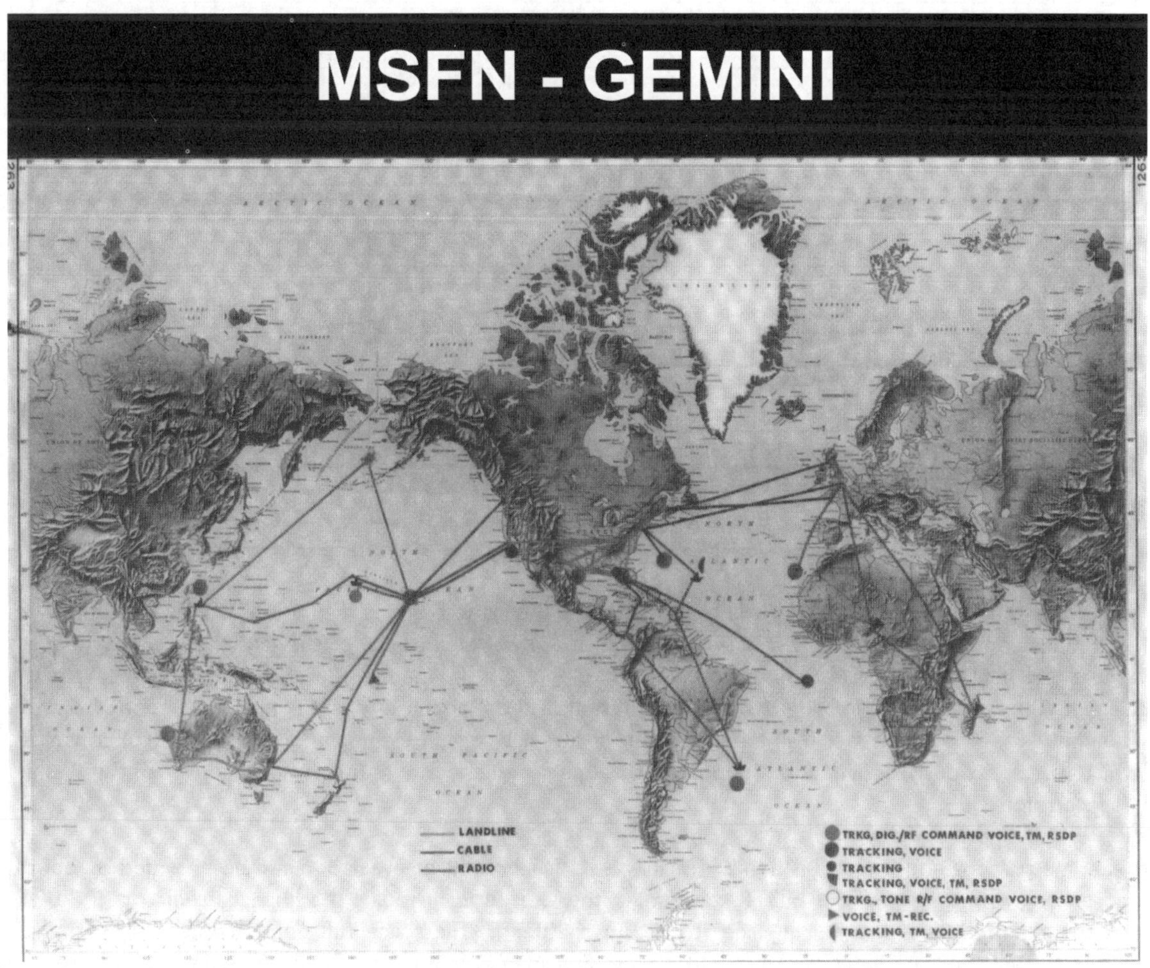

FIG. 24

TABLE III
NETWORK CAPABILITIES

	Tracking				Telemetry										
	Radar		Mistram	Acq. aid	GLV	Agena		Spacecraft				Command		A/G voice	
	C	S	or other as listed		PCM	R/T	D/T	Links received	R/T	D/T	R.S.D.P.	DCS	Tone	UHF	HF
CNV	X	X	GE Mod III	2	X	X[b]	X	3	X	X	X	X[c]	X[c]	X[d]	X[d]
GBI	X				X[a]	X[b]	X	3	X[b]	X	X	X[c]	X[e]	X[d]	X[d]
GTI	X					X[b]	X	3	X[b]	X	X	X[c]	X[e]	X[d]	X[d]
ANT	X			2		X[b]	X	3	X[b]	X	X	X[c]	X[e]	X[d]	X[d]
ASC	X					X[a]		2	X[a]					X[d]	X[d]
VAL			X												
ELU			X												
BDA	X	X		2		X[b]	X	3	X[b]	X	X	X[c]		X[d]	X[d]
CYI	X	X		2		X[b]	X	3	X	X	X	X		X	X
KNO				X		X[a]		2	X[a]					X[d]	X[d]
TAN				X		X[a]		2	X[a]					X[d]	X[d]
CRO	X	X		2		X[b]	X	3	X	X	X	X		X	X
CTN				X		X[a]		2	X[a]					X[d]	X[d]
HAW	X	X		2		X[b]	X	3	X	X	X	X		X	X
GYM		X		2		X[b]	X	3	X	X	X			X	X
TEX		X		2		X[b]	X	3	X	X	X	X[c]		X	X
RKV				2		X[b]	X	3	X	X	X	X		X	X
CSQ				2		X[b]	X	3	X	X	X	X		X	X
RTK	X			X		X[a]		2	X[a]					X[d]	X[d]
CAL	X			X										X	X
WHS	X			X											
EGL	X			X										X[c]	X[c]
A/C								2	X[a]	X[a]					
PRE	X														

a. Record only b. Remoted to MCC-H c. Remoted from MCC-H
d. Remoted to and from MCC-H e. Remoted from MCC-H and CNV.

Support for the Gemini VII mission will be provided by the same basic configuration of the GOSS that was established for support of the Gemini-Agena mission; however, in light of the fact that certain Gemini-peculiar elements of the GOSS will be reserved for support of the Gemini VII mission plus the fact that the Gemini VII mission will be in progress for the duration of the Gemini VI mission, minor deviations from the Gemini VI GOSS configuration will be effected in order that certain elements (i.e., flight control consoles), which were previously designed for support of the Agena vehicle, can be utilized for support of the Gemini 6 spacecraft.

Real-time tracking and the acquisition of tracking data for post-flight evaluation will be provided by optical and photographic systems, the MISTRAM system, the G.E. Mod III radar, the Impact Predictor (IP 7094), and the MSFN C-Band radars. In order to accurately identify the spacecraft being tracked, the Gemini 6 spacecraft will use the Gemini C-Band Radar Pulse Spacing Code and the Gemini 7 spacecraft will use the Agena vehicle Pulse Spacing Code.

PCM Telemetry will be utilized for relaying the status of the astronauts and the onboard instrumentation from the spacecraft to the GOSS elements. The PCM data which is received by those MSFN sites which are manned by flight controllers will be displayed directly at the receiving site. PCM data received by "real-time" MSFN sites (unmanned) will be transmitted in real time to the MCC-H for display. In order to facilitate adequate differentiation of data when telemetry is being received from both the Gemini 6 and the Gemini 7 spacecraft by a single site, a different transmitter on each spacecraft will be used (i.e., real-time vs. standby) and, as a result, the data will then be separated on a frequency basis.

During the launch phase of the mission, a tone command system at MCC-H will be employed for transmitting ASCO (Auxiliary Second Stage Cut-off), a tone command system at Cape Kennedy will be utilized, as required, for transmitting range safety commands (i.e., Arm, Destruct), and a Digital Command System (DCS) at the Mission Control Center, Cape Kennedy, will be used for transmitting spacecraft Inertial Guidance System updating commands. Direct ground control of certain spacecraft systems will be accomplished during the orbital phase of the mission via the DCS's located at five of the MSFN sites and the Master Digital Command System at MCC-H. It should be noted at this point that the binary structures of the Gemini 6 command words are identical to those of the Gemini 7 command words. Consequently, in order to avoid undesirable, simultaneous transmission of commands to both spacecraft in those instances when both spacecraft are within range of a single transmitting antenna, it will be necessary for the astronauts of the spacecraft, for which commands are not intended, to turn off the on-board command receiver.

Voice Communications with the spacecraft will be provided using HF (15.016 mc) and UHF (296.8 mc) frequencies. Those sites which will be manned by flight controllers have the capabilities for communication directly with the spacecraft, but the voice facilities at certain of the non-flight controller sites will be remotely controlled from the MCC-H. If the Gemini 6 and Gemini 7 spacecraft are within range of a single site, simultaneous voice reception from both spacecraft is provided for by site instrumentation, regardless of the distance between the two spacecraft; however, ground equipment constraints restrict simultaneous voice transmission to the spacecraft to those periods when the distance between the two spacecraft is 20° radar look angle or less. The spacecraft toward which voice communications will be directed, when the distance between the two spacecraft is greater than 20°, will be procedurally selected.

The accurate and timely flow of information between the various GOSS elements is provided by the NASA Communication Network (NASCOM) utilizing land-line, cable, microwave link, and HF single-sideband circuits in conjunction with appropriate switching centers. All HF communications and communications interconnecting critical components of the Mission Control and Launch Instrumentation Systems have redundant circuits if required. The tracking ship Coastal Sentry Quebec (CSQ) will use SYNCOM III as backup for HF communications. Communications capabilities provided include: (1) Voice communications for mission operations, maintenance, and coordination, (2) teletype communications within MSFN to provide for summary, command, and acquisition messages, and telemetry and radar data summaries; (3) high speed and wide-band data communications for transmission of real-time radar, command, telemetry, and display data as well as certain delayed-time data; and (4) closed circuit TV between the MCC-H and the launch area during launch operations.

Communications used by the Atlantic and Pacific Recovery Forces will employ HF single-sideband and UHF radio links as well as land lines and underwater cables. Relay aircraft will communicate between the airborne and surface elements of the Recovery Forces. The DOD communications net, which supports the DOD Recovery Forces, interfaces with NASCOM at Kunia, Cape Kennedy, and Adelaide.

NOMINAL MISSION PLAN

Gemini VII will be launched (Figure 25) from Cape Kennedy Launch Complex 19 at 2:30 p.m. EST at a launch azimuth of 083.6° into an 87-183 nautical mile elliptical orbit having an inclination of 28.87°. At SECO+30 seconds the spacecraft will separate from the booster using a separation velocity of two feet per second. Immediately after separation the spacecraft will turn around (to blunt end forward) and perform a station keeping exercise on the booster second stage. This operation will continue until approximately five minutes after entering the first darkness period (about twenty-five minutes Ground Elapsed Time (GET) from liftoff). Following this activity, the D-4 (Celestial Radiometry) experiment will be conducted without an Earth background during the remainder of the first darkness period.

At the third spacecraft apogee (GET = 3 hours, 50 minutes) the perigee will be raised to 108 nautical miles in order to establish a spacecraft orbital lifetime of fifteen days. The next several days will be devoted to conducting the assigned experiments.

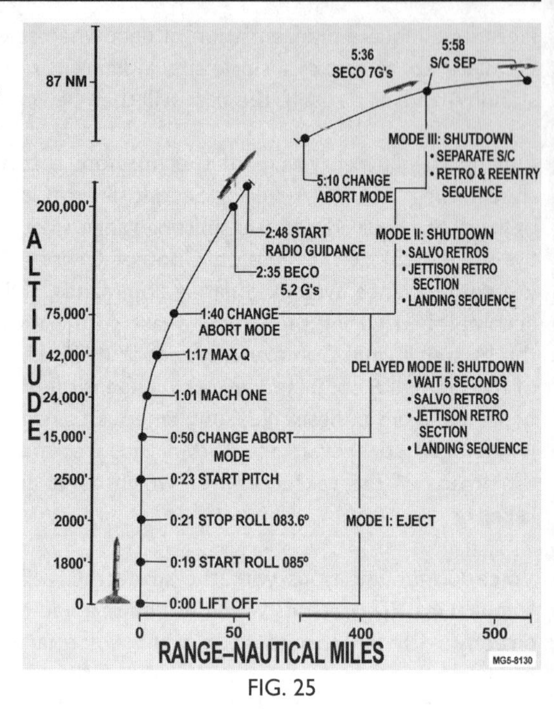

FIG. 25

The Gemini 7 orbit will be circularized to 161 nautical miles to provide the proper target orbit for the Gemini 6 spacecraft. The circularization maneuver depends on the decay rate of the Gemini 7 orbit and the expected liftoff time of Gemini 6. Nominally, after approximately 120 hours elapsed time a posigrade thrust resulting in a velocity increment of approximately one hundred feet per second will be applied at apogee to circularize the spacecraft orbit to 161 nautical miles.

Gemini 6 nominal liftoff time is planned to occur eight days, nineteen hours and eight minutes after Gemini 7 liftoff. At orbital insertion, spacecraft 6 will trail spacecraft 7 by approximately 1050 nautical miles. Rendezvous should occur approximately five hours and thirty-six minutes after Gemini 6 liftoff. Following initial rendezvous the Gemini 6 will take photographs of the Gemini 7 aft-firing thrusters, and perform a station-keeping exercise. The Gemini 6 will station keep on the Gemini 7 for two full revolutions. Gemini 7 will station keep on Gemini 6 for the daylight portion of one revolution. Following the station keeping exercise, the Gemini 6 will conduct the assigned experiments and position itself for reentry. Gemini 7 will photograph the Gemini 6 retrofire and as much of its reentry as possible. The rendezvous portion of the combined Gemini VII / Gemini VI-A mission is described in detail in Gemini VI-A Mission Operation Report (No. M-913-65-08).

The remaining experiment activities scheduled for Gemini VII will be completed following completion of the rendezvous activities. Gemini 7 will retrofire near the end of the 206th revolution and splash down will occur in the West Atlantic recovery zone at the beginning of the 207th revolution.

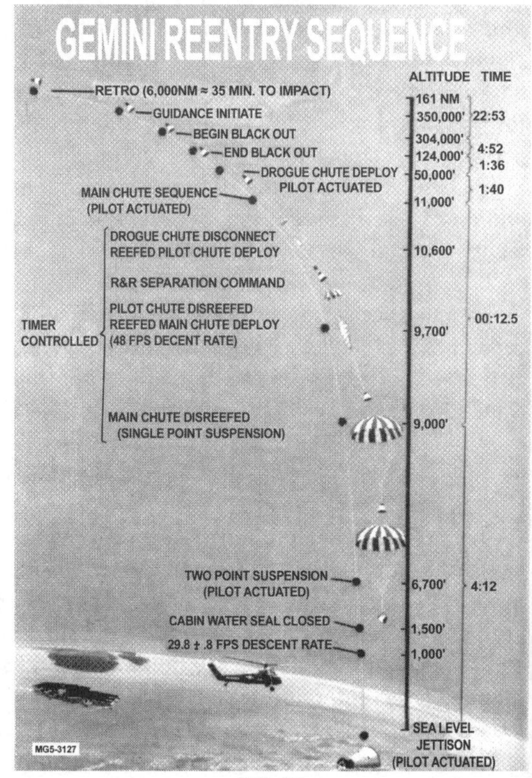

FIG. 26

Gemini VII will terminate in the West Atlantic Recovery zone at the beginning of revolution 207. Spacecraft landing should occur about 35 minutes after retrofire over Canton Island. An aiming point to place the spacecraft about 360 nautical miles south of Bermuda will be used; however, its location will vary with actual retrofire position, reentry parameters, and orbital altitude. The reentry sequences are illustrated in Figure 26.

As with earlier missions, the recovery forces consist of launch site small craft and amphibious vehicle (Figure 27) which will support a 15 mile strip from Complex 19 to roughly 41 miles seaward and three miles inland to the Banana River. Continuing eastward are the launch abort forces made up of a Navy Fleet Tug, an Aircraft Carrier and Oiler, three Destroyers and six long range aircraft. Figure 28 illustrates the positioning of the launch abort

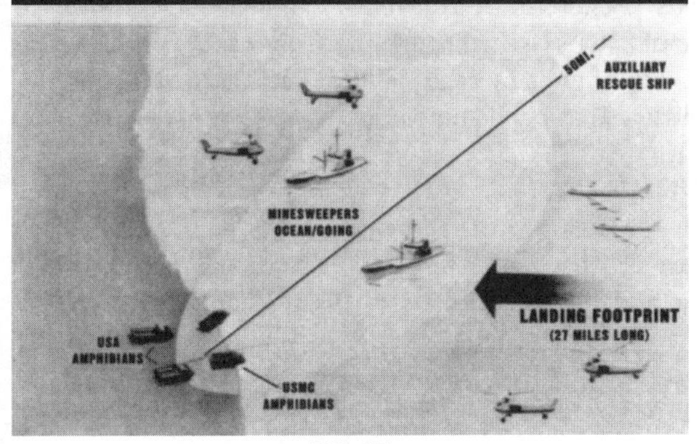

FIG. 27

recovery forces and Figure 29 illustrates the launch abort sequences. These units are prepositioned to provide recovery support should the need arise for an abort prior to spacecraft insertion. Medical assistance is available in each instance. Following insertion, all units will deploy in and around the recovery zones: West Atlantic, Mid Atlantic and West and Mid Pacific. This insures proper coverage should a planned landing area be required as a result of an unexpected reentry. In addition to the surface units, there are groups of long range aircraft, equipped with drop kits and pararescue / paramedic personnel, located throughout the world at contingency staging bases. They would be used in case support could not be provided by the surface units within the four landing zones.

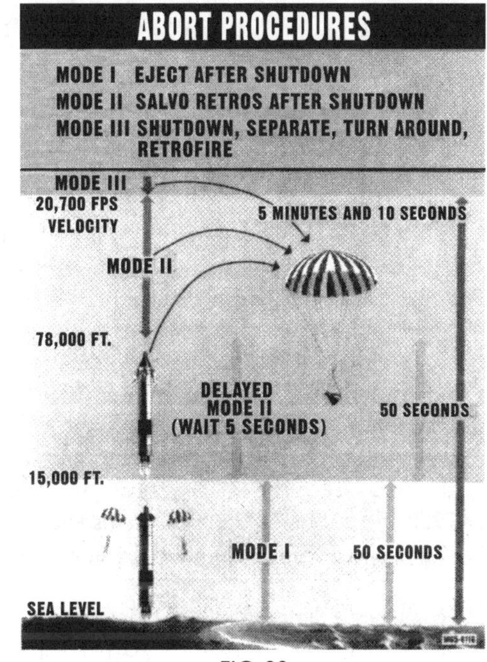

FIG. 29

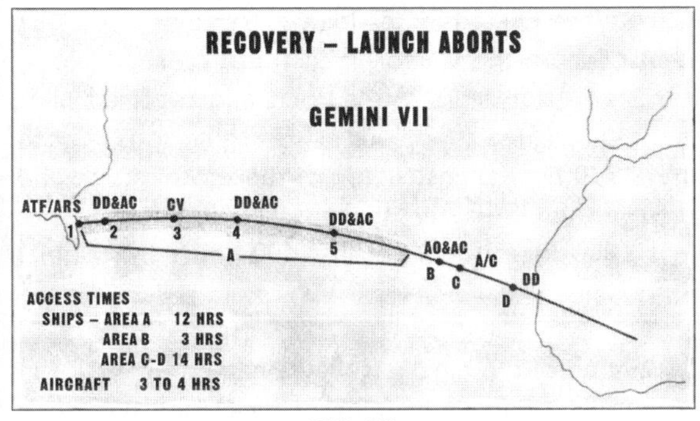

FIG. 28

The primary recovery unit will be the Aircraft Carrier since greater flexibility is available with its more extensive medical facilities and aircraft capability. Should it be necessary to use the Oiler or one of the Destroyers, however, the proper equipment for spacecraft handling and medical needs is installed, and crews are trained and ready.

Primary and Secondary Landing Zones and the location of contingency staging bases are shown in Figure 30.

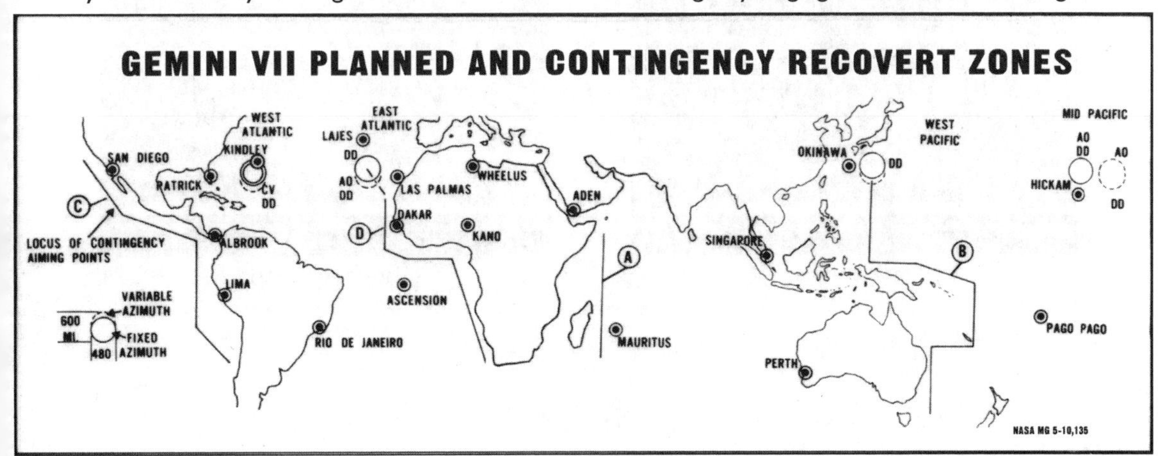

FIG. 30

Recovery coordination will be conducted in identical fashion to the previous missions. All units are under the operational control of the Defense Department Manager for Manned Space Flight at the Recovery Control Center in Houston. Control of Atlantic units is exercised through the Commander, Task Force 140, located at Cape Kennedy. Similarly, Pacific control is exercised through the Commander, Task Force 130, at Kunia, Hawaii. Communication is maintained with all units by each Task Force Commander and the DOD Manager from T-4 hours until the spacecraft retrieval has been completed.

ASTRONAUTS

The Command Pilot will be Frank Borman and Pilot will be James A. Lovell. The backup flight crew consists of Edward H. White II as Command Pilot and Michael Collins as Pilot. Their pictures and biographies follow.

FRANK BORMAN

Born in Gary, Indiana on March 14, 1928, Borman is married to the former Susan Bugbee of Tucson, Arizona and has two sons. He received his B.S. degree from the United States Military Academy at West Point in 1950 and his M.S. degree in aeronautical engineering from the California Institute of Technology, Pasadena, in 1957. Borman entered the Air Force after graduation from West Point and received pilot training at Williams AFB, Arizona. From 1951 to 1956 he was assigned to various fighter squadrons in both the United States and the Philippines and from 1957 until 1960 he was instructor of thermodynamics and fluid mechanics at the Military Academy. He graduated from the USAF Aerospace Research Pilots School in 1960 and remained there as an instructor until 1962. He has logged more than 4,400 hours flying time, including more than 3,600 hours in jet aircraft.

He is a United States Air Force Lieutenant Colonel and was one of the nine astronauts named by NASA in September 1962. In addition to participating in the overall astronaut training program, he has been assigned a variety of special duties. He was command pilot of the backup crew for the second Gemini manned space flight.

Frank Borman

James A. Lovell

JAMES A. LOVELL

Born in Cleveland, Ohio on March 25, 1928, Lovell is married to the former Marilyn Gerlach of Milwaukee, Wisconsin and has three children. He received his B.S. degree from the United States Naval Academy in 1952. His last Navy assignment was as flight instructor and safety officer with Fighter Squadron 101 at the Naval Air Station at Oceana, Virginia. From January 1958 until July 1961, he was test pilot at the Navy Air Test Center, Patuxent River, Maryland. His work there included service as program manager for the F4H Weapon System Evaluation. He graduated from the Aviation Safety School of the University of Southern California in 1961. He has logged more than 3,000 hours flying time, including more than 2,000 hours in jet aircraft.

He is a United States Navy Commander and was one of the nine astronauts named by NASA in September 1962. In addition to participating in the overall astronaut training program, he has been assigned special duties, which include monitoring design and development of recovery and crew life support systems. These include spacesuits, environmental control systems and developing techniques for lunar and Earth landing and recovery. He was pilot of the backup crew for the second Gemini manned flight.

EDWARD H. WHITE II

Born in San Antonio, Texas on November 14, 1930, White is married to the former Patricia Finegan of Washington, D.C. and has two children. He received his B.S. degree from the United States Military Academy at West Point in 1952 and his M.S. degree in aeronautical engineering from the University of Michigan in 1959. He served as an experimental test pilot with the Aeronautical Systems Division at Wright-Patterson Air Force Base. In this assignment, he made flight tests for research and weapons systems development, wrote technical engineering reports, and made recommendations for improvement in aircraft design and construction. He has logged more than 3,600 hours flying time, including more than 2,200 hours in jet aircraft.

He is a United States Air Force Lieutenant Colonel and was one of the nine astronauts named by NASA in September 1962. White was pilot of the second manned mission of the Gemini Program, Gemini IV, in June 1965. The flight was 97 hours and 59 minutes in duration and he performed 22 minutes of Extravehicular Activity, and in the process became the first man to maneuver himself in space by means of a hand-held, self contained maneuvering device. White has been awarded the NASA Distinguished Service Medal and the Astronaut Wings.

Edward H. White, II

Michael Collins

MICHAEL COLLINS

Born in Rome, Italy on October 31, 1930, Collins is married to the former Patricia Finnegan of Boston, Massachusetts and has three children. He received his B.S. degree from the United States Military Academy at West Point. Collins entered the Air Force after graduation. He served as an experimental flight test officer at the Air Force Flight Test Center, Edwards AFB, California. In that capacity he tested performance and stability and control characteristics of Air Force aircraft, primarily jet fighters. He has logged more than 3,000 hours flying time, including more than 2,700 hours in jet aircraft.

He is a United States Air Force Major and was one of the third group of astronauts named by NASA in October 1963. In addition to participation in the overall astronaut training program, he has been assigned special duties, which include pressure suits and extravehicular experiments.

ASTRONAUT ACTIVITIES

PRE-MISSION:

The Gemini VII flight crew was selected July 1, 1965 and began their concentrated mission training immediately. In addition to the extensive general training received prior to flight assignment, such as familiarization with high accelerations, zero gravity, and various survival techniques, the following mission-specific preparations have or will be accomplished prior to launch:

- Familiarization with launch, launch abort, and reentry profile for the Gemini VII mission.
- Egress and recovery activities using a spacecraft boilerplate and actual recovery equipment and personnel.
- Celestial pattern recognition at the Moorehead Planetarium.
- Parachute descent training over land and water.
- Suit, seat, and harness fittings.
- Training sessions in the Gemini Mission Simulators.
- Detailed systems briefings; detailed experiment briefings; Flight Plan and Mission Rules Reviews.
- Participation in Mock-up Reviews, SEDR Reviews, Subsystem Tests, Spacecraft Acceptance Review and Flight Readiness Reviews,

In final preparation for flight, the crew participated in network launch abort simulations, Joint Combined Systems Test, and the final Simulated Flight Test. At F-2 days, the major flight crew medical examination is administered to determine readiness for flight and obtain data for comparison with post flight medical examination results. The primary crew is supported at all times.

MISSION:

At ignition, the crew begins their primary launch phase task of assessing system status and detecting abort situations. Thirty seconds after SECO, the Command Pilot initiates forward thrusting and the Pilot actuates spacecraft separation. Ground computations of insertion velocity corrections are received and implemented by forward or aft thrusting. After successful insertion and completion of the insertion check list, the detailed Flight Plan is commenced.

The detailed Flight Plan consists mainly of activities which are dependent on several variables such as position on ground track, daylight, weather, and attitude control fuel. Therefore, the Flight Plan is actually a departure point for real time modifications to the activities schedule as these variables become determined. The basic Flight Plan is discussed in the Mission Plan section of this report.

POST-MISSION:

Following landing the flight crew will participate in the scheduled activities listed in Table IV.

TABLE IV
POST FLIGHT ACTIVITIES

Day	Activity
R	Medical exam
R+1	Medical exam and Technical debriefing
R+2, 3, 4	Technical debriefing
R+5	Management and Project debriefing
R+6	Debriefing with follow-on flight crews and photo identification
R+7	Work on Pilot's report
R+8	Systems debriefing
R+9	Scientific debriefing and work on Pilot's report

R = Recovery

ASTRONAUT EQUIPMENT

The spacesuit, shown in Figure 31 (in both partial wear and fully donned states) and designated as the G-5C, will be worn by the Gemini VII crew and has been developed to meet the basic design objectives of reduced bulk and weight and improved unpressurized comfort and mobility. At the same time, it provides satisfactory pressurized mobility to control the spacecraft for earth reentry in the event this proves necessary.

The soft helmet with the internally worn crash helmet provides for easier stowage in the partial wear mode. The elimination of the large rotating neck bearing and helmet tie-down system not only provides additional comfort and clears up frontal suit areas, but also permits quicker emergency donning. The capability of removing the outer jump boots without reducing suit pressure integrity also contributes to long-term wear comfort. The pliability of the suit allows easier waste management and accessibility to spacecraft remote stowage areas. The elimination of excess bulk makes possible complete removal of the suit within the confines of the spacecraft. During Gemini VII mission the suits will be removed by the crew and redonned later in the mission.

FIG. 31

MENU

The Menu for this mission consists of bite-size and reconstituted food and is depicted in Table V.

TABLE V

Menu I Day 1-5-9-13	Menu II Day 2-6-10-14	Menu III Day 3-7-11	Menu IV Day 4-8-12
Meal A Grapefruit drink Apricot cereal cubes (8) Sausage patties (2) Banana pudding Fruit cocktail	**Meal A** Grapefruit drink Chicken and gravy Beef sandwiches (6) Applesauce Peanut cubes (6)	**Meal A** Salmon salad Green peas Bread cubes (8) day 3 only Gingerbread (6) Cocoa	**Meal A** Strawberry cereal cubes* Bacon squares (4) Ham and applesauce Chocolate pudding Orange Drink
Meal B Beef and vegetables Potato salad Cheese sandwiches (6) Strawberry cubes (6) Orange drink	**Meal B** Orange-grapefruit drink Beef pot roast Bacon and egg bites (6) Chocolate pudding	**Meal B** Grapefruit drink Bacon squares (4) Chicken and vegetables Apricot cubes (6) Pineapple fruitcake (6)	**Meal B** Beef and gravy Corn chowder Brownies (6) Peaches
Meal C Orange-grapefruit drink Tuna salad Apricot pudding Date Fruitcake (4)	**Meal C** Potato soup Shrimp cocktail Date fruitcake (4) Orange Drink	**Meal C** Spaghetti and meat Cheese sandwich (6) Butterscotch pudding Orange Drink	**Meal C** Coconut cubes (6) Cinnamon toast (6) Chicken salad Grapefruit drink

* (8) day 8 only

PROJECT COST AND ALLOCATION OF FUNDS

($ Millions)

	FY 62	FY 63	FY 64	FY 65	FY 66	FY 67
Spacecraft	30.3	205.1	280.5	165.3	122.7	19.1
Launch Vehicle	24.4	79.1	122.7	115.4	88.6	8.5
Operational Support	0.1	4.9	15.7	27.7	30.8	13.0
RD&O	54.8	289.1	418.9	308.4	242.1	40.6

Total expected project cost based upon obligational authority and allocation:

Spacecraft	$ 823.0
Launch Vehicle	438.7
Support	92.2
Total RD&O	$ 1353.9

MISSION MANAGEMENT RESPONSIBILITY

Management of the Gemini Program is conducted by the Gemini Program Director who exercises control through project management at the Manned Spacecraft Center. Each mission is conducted by a Mission Director who acts in behalf of the Associate Administrator for Manned Space Flight from spacecraft commitment to flight test until the mission is completed.

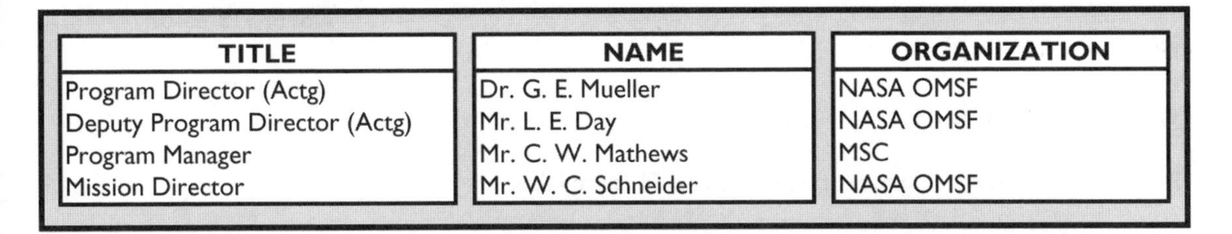

TITLE	NAME	ORGANIZATION
Program Director (Actg)	Dr. G. E. Mueller	NASA OMSF
Deputy Program Director (Actg)	Mr. L. E. Day	NASA OMSF
Program Manager	Mr. C. W. Mathews	MSC
Mission Director	Mr. W. C. Schneider	NASA OMSF

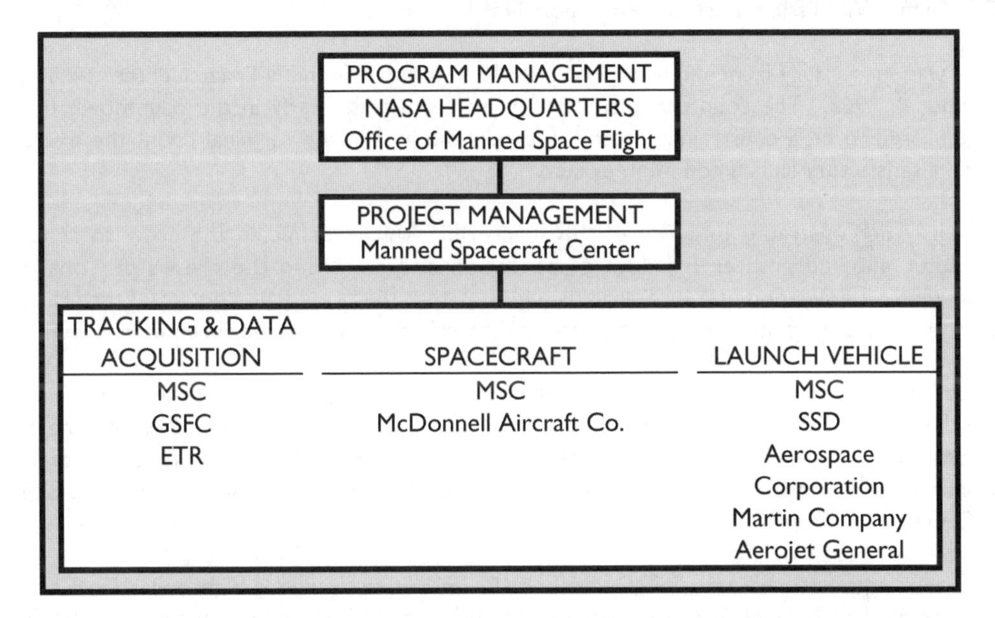

OPERATIONS ORGANIZATION FOR GEMINI VI-A MISSION

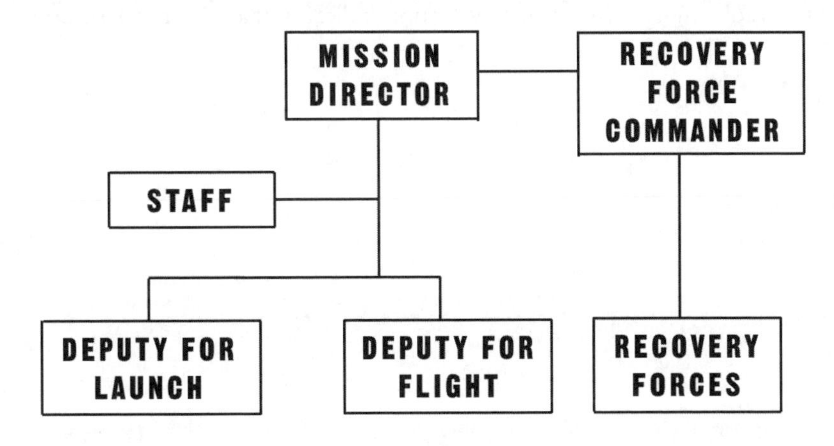

FIG. 32

NATIONAL AERONAUTICS AND SPACE ADMINISTRATION
WASHINGTON, D.C. 20546

Mission Operations Report
No. M-913-65-07

MEMORANDUM

December 28, 1965

TO: A / Administrator

FROM: M / Associate Administrator for Manned Space Flight

SUBJECT: Gemini VII Mission, Post Launch Report No. 1

Gemini VII was successfully launched from Complex 19 at John F. Kennedy Space Center at 2:30 p.m. EST on December 4, 1965. The countdown and launch were flawless. Early in the countdown the adverse weather appeared to be a constraining factor, but near the end of the terminal count the overcast skies cleared and a satisfactory launch condition existed.

The previously demonstrated integrity of the spacecraft, which permitted the crew to remove their pressure suits, allowed greater freedom of operation and enhanced the crew's personal comfort. Preliminary medical evaluation and data analysis indicates that the crew is in very good medical condition and the effects of exposing the crew to long periods of weightlessness have not been detrimental.

The fuel cell demonstrated excellent electrical characteristics throughout the mission. The intermittent lighting of the fuel cell differential pressure light did cause some concern but did not, at any time, become a mission constraint. The tremendous success of the Gemini VI-A and VII missions has again confirmed the operational capability of the launch and space vehicles, the Manned Spaceflight Network, and the Mission Control Center.

Gemini VII terminated at 9:05 a.m. EST December 18, 1965, approximately eight miles from the planned recovery zone after completing 206 revolutions in 330 hours and 35 minutes. Recovery was accomplished swiftly and effectively in the West Atlantic by the aircraft carrier U.S.S. WASP.

The primary objective to demonstrate man's capability to perform complex functions throughout the 14-day period was successfully met. Preliminary evaluation indicates that the crew returned in excellent physical condition.

George E. Mueller

Enclosure:
MOR No. M-913-65-07
Post Launch Report No. 1

GEMINI VII MISSION
POST LAUNCH REPORT NO. I

COUNTDOWN

The final countdown began at 9:30 a.m. EST and proceeded without a flaw or hold and liftoff occurred at 2:30 p.m. EST December 4, 1965, as scheduled. Early in the countdown the adverse weather appeared to be a constraining factor, but near the end of the terminal count the overcast skies cleared and a satisfactory launch condition existed.

LAUNCH

FIG. I

The launch and near perfect insertion were flawless, with all systems operating nominally and remaining in this configuration throughout the launch phase. The crew reported an unexpected indication in the section two fuel cell reactant supply system differential pressure between 170 and 185 seconds after liftoff, but resolution was deferred until after insertion. The launch vehicle performed satisfactorily with no apparent anomalies. The station-keeping exercise with the booster, immediately following spacecraft separation, was accomplished at a crew-reported separation distance of 50 feet. Launch and insertion parameters are shown in Table I and launch is depicted in Figure I.

TABLE I

Parameter	Planned	Actual
Insertion Velocity, ft/sec	25,804	25,793
Flight Path Angle at Insertion, °	0.01	0.06
Insertion Altitude, N.M.	87.1	87.2
Inclination Angle, degrees	28.87	28.89
Period, Minutes	89.5	89.39
Apogee, N.M.	183.1	177.1
Perigee, N.M.	87.1	87.2
BECO, seconds	155.35	155.61
SECO, seconds	338.61	337.01

MISSION

Considering the fourteen-day vacuum soak the spacecraft endured, it performed remarkably well. There were some anomalies which arose and will be further discussed, but no problems came up that seriously endangered the success of the mission.

The first anomaly was the intermittent lighting of the fuel cell differential pressure light. This warning light indicates the pressure differential between the fuel cell oxygen and the fuel cell product water. The anomaly continued throughout the mission but was not considered to be a constraining item. Several times during the mission the fuel cell stacks indicated a variation in load sharing and showed definite signs of degradation. The fuel cells, as designed, have a capability of series and parallel switching of the stacks and consequently through manipulation of the switches the fuel cell continued to meet all demands placed upon it.

The Reactant Supply System oxygen pressure was low during the early phase of the flight but the condition was corrected by supplying oxygen from the Environmental Control System by means of a cross feed valve. This valve had been installed to preclude a repeat of a similar anomaly which had existed during Gemini V. This action corrected the pressure and the discrepancy did not reappear.

The telemetry tape dump recorder utilized for storing and dumping information in delayed time became inoperative after approximately 200 hours of flight, and numerous attempts to alleviate the condition were unsuccessful.

Experiment M-8, "In-Flight Sleep Analysis," was terminated after 55 hours and 15 minutes GET because one of the electroencephalograph (EEG) sensors pulled loose and the crew was unsuccessful in replacing the sensor. The photometer used for experiment D-5, "Star Occultation Navigation," was determined to be susceptible to radio frequency interference at a frequency of approximately 200 megacycles. At this frequency, the instrument's amplifier became saturated and control using the gain wheel and calibration button was lost; therefore, activity using the D-5 equipment was curtailed. The failure of the telemetry tape dump recorder and the susceptibility of the photometer to radio frequency interference resulted in the termination of experiment MSC-12, "Landmark Contrast Measurements."

Prior to the launch of Gemini VI-A, spacecraft 7 was maneuvered to establish a near-circular 159 by 163 nautical mile target orbit. Spacecraft 6 acquired radar lock-on with spacecraft 7 four hours and sixteen minutes after liftoff and at a distance of 235 miles. The Gemini VII acquisition aid lights were observed by the Gemini VI-A crew throughout the terminal phase of the rendezvous. Gemini VII was not maneuvered during the rendezvous but Command Pilot Borman controlled spacecraft attitude to maintain radar lock.

Rendezvous activities with spacecraft 6 on the eleventh day of the Gemini VII mission included station-keeping, fly-arounds, in-plane and out-of-plane maneuvers, photographing thruster plumes, and visual inspection of spacecraft 7. In order to conserve fuel, Gemini VII performed "station keeping" exercises for only 40 minutes of the "station keeping" period. Following five hours and nineteen minutes of rendezvous activities, the two spacecraft separated to a variable distance of 8-32 nautical miles for the sleep period and the remainder of Gemini VI's mission. On the twelfth day of the Gemini VII mission, spacecraft 6 returned to earth and spacecraft 7 continued its mission.

At approximately 267 hours elapsed time (during rendezvous activities), the crew reported substantial amounts of water being discharged from the suit inlet hoses. It was hypothesized that the cause was either excessive condensation as a result of a reduction in cabin temperature, or backup of condensate water as a result of the launch cooling heat exchanger being full. Corrective action was taken for both possible causes. The suit fans were turned on to increase evaporation and the spacecraft was rolled at 10 degrees per second to aid in forcing out the excess water from the launch cooling heat exchanger. The crew of spacecraft 6 observed the water venting and the condition was alleviated.

The orbital-attitude yaw-right thruster chambers #3 and #4 became partially inoperative after approximately 281 hours elapsed time. Corrective action was taken to provide yaw-right impulse by use of the orbital maneuver right-side forward firing thruster. The time of the occurrence of the anomaly and corrective action was such that the crew was unable to observe spacecraft 6 retrofire.

Valuable data was obtained on all twenty experiments. The complete success of each, however, will not be known until post-flight analysis allows comparison of actual with expected results. Some of the primary anomalies contributing to incomplete performance have already been discussed: oversaturation of the photometer amplifier, removal of the EEG sensors, and the inoperable telemetry tape dump recorder. Obvious problems arose due to weather over the two week period in which the photography and visual recognition experiments were affected. For example, the Laser experiment (MSC-4) was attempted repeatedly; however, the crew acquired the beam only twice and in both instances were unable to modulate it. Had visibility been more optimum, the probability of successful transmission to ground would have been greater.

The medical experiments were conducted continuously throughout the mission, and appear to have been a success – aside from the In-Flight Sleep Analysis discussed before. Similarly, the other experiments, scientific, engineering and technological, show preliminary success.

Gemini VII terminated at 9:05 a.m. EST December 18, 1965, approximately 8 nautical miles from the primary recovery ship after completing 206 revolutions in an elapsed time of 330 hours and 35 minutes. Recovery was accomplished swiftly and effectively in the West Atlantic by the aircraft carrier U.S.S. WASP. Table II lists the major recovery events.

TABLE II

Time, GMT	Event
1358	Aircraft carrier U.S.S. WASP had spacecraft on radar
1404	Visual sighting by recovery aircraft
1405	Spacecraft landing
1407	Spacecraft / Aircraft communications established
1417	Floatation collar attached and inflated
1424	Astronauts in life raft
1426	Astronauts aboard the helicopter
1437	Astronauts aboard the carrier
1510	Spacecraft retrieved and secured aboard the carrier

Crew response during the mission was accurate, sharp, and punctuated with good humor. Their attitude and cooperation remained of the highest quality. Each crewman consumed approximately 2000 calories per day and more than four pounds of water per day. Preliminary post-flight evaluation indicates both crew members are in excellent physical condition.

The flight crew participated in the scheduled post-flight activities listed in Table III.

TABLE III

12/18/65	Medical examination
12/19/65	Depart carrier for KSC
12/20-21/65	Technical Debriefing
12/22/65	Depart KSC for MSC
12/23/65	Management Debriefing
12/24-25/65	Christmas Vacation
12/26/65	Pilot's Report
12/27/65	Debriefing with follow-on flight crews and photo identification
12/28/65	Systems Debriefing
12/29/65	Scientific Debriefing
12/30/65	Press Conference

PROJECT GEMINI DRAWINGS AND TECHNICAL DIAGRAMS

GEMINI SPACECRAFT PARACHUTE LANDING SYSTEM

from Gemini Familiarization Manual

revision 31 December 1964

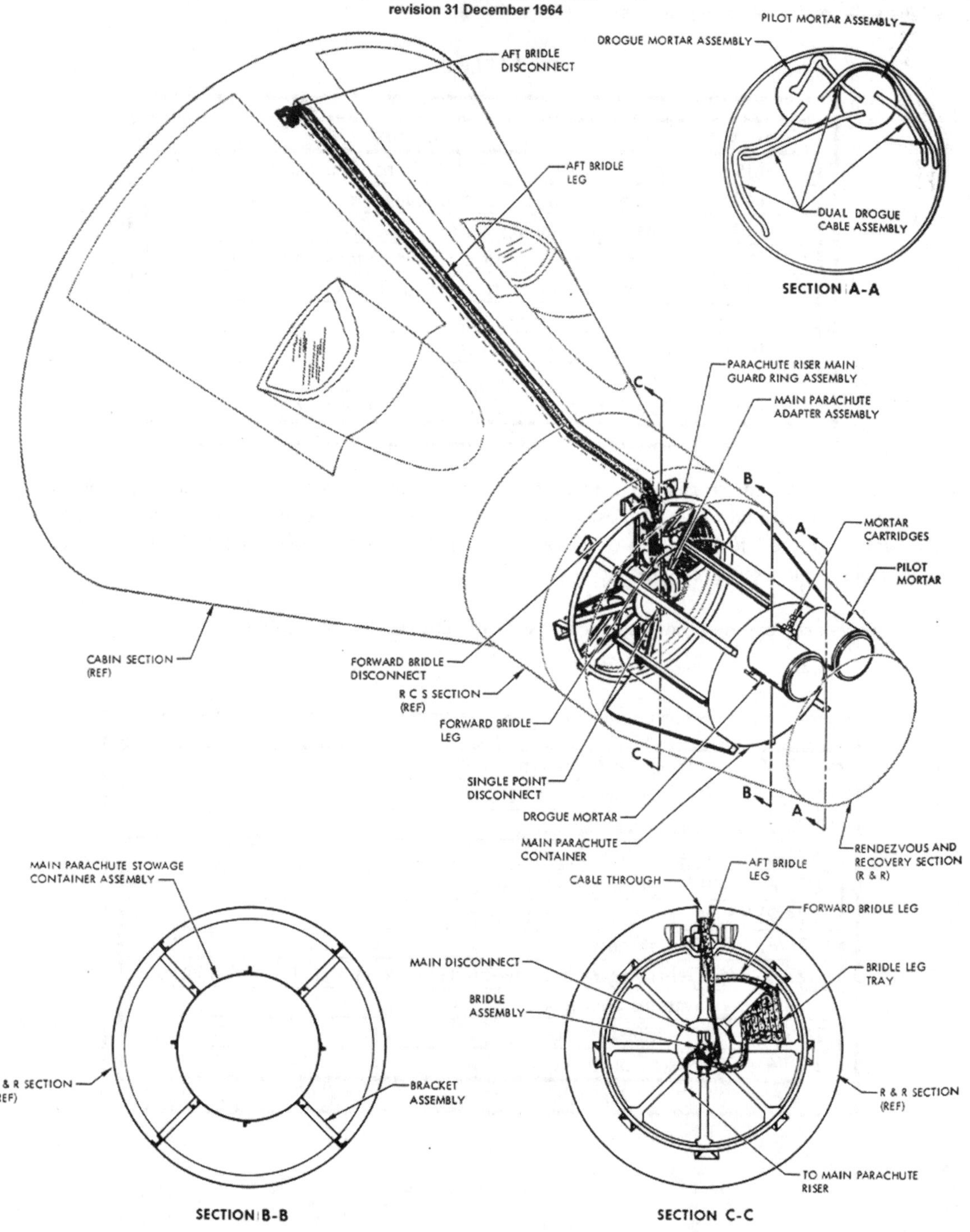

PILOT MORTAR ASSEMBLY
DROGUE MORTAR ASSEMBLY
DUAL DROGUE CABLE ASSEMBLY
SECTION A-A
AFT BRIDLE DISCONNECT
AFT BRIDLE LEG
PARACHUTE RISER MAIN GUARD RING ASSEMBLY
MAIN PARACHUTE ADAPTER ASSEMBLY
MORTAR CARTRIDGES
PILOT MORTAR
CABIN SECTION (REF)
FORWARD BRIDLE DISCONNECT
R C S SECTION (REF)
FORWARD BRIDLE LEG
SINGLE POINT DISCONNECT
DROGUE MORTAR
MAIN PARACHUTE CONTAINER
RENDEZVOUS AND RECOVERY SECTION (R & R)
MAIN PARACHUTE STOWAGE CONTAINER ASSEMBLY
R & R SECTION (REF)
BRACKET ASSEMBLY
SECTION B-B
CABLE THROUGH
AFT BRIDLE LEG
FORWARD BRIDLE LEG
MAIN DISCONNECT
BRIDLE ASSEMBLY
BRIDLE LEG TRAY
R & R SECTION (REF)
TO MAIN PARACHUTE RISER
SECTION C-C

GEMINI EQUIPMENT ARRANGEMENT

from Press Reference Book for
Gemini Spacecraft Number 11
REVISION 30 AUGUST 1966

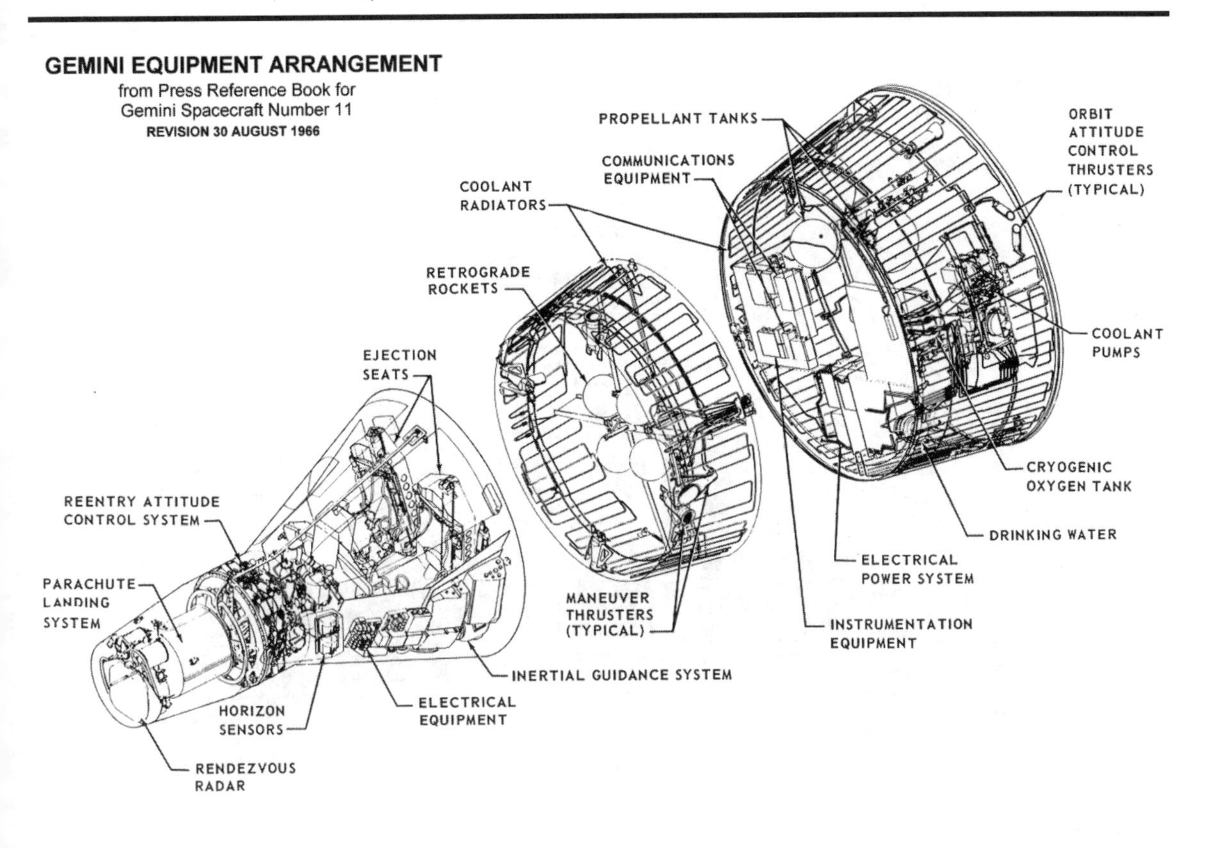

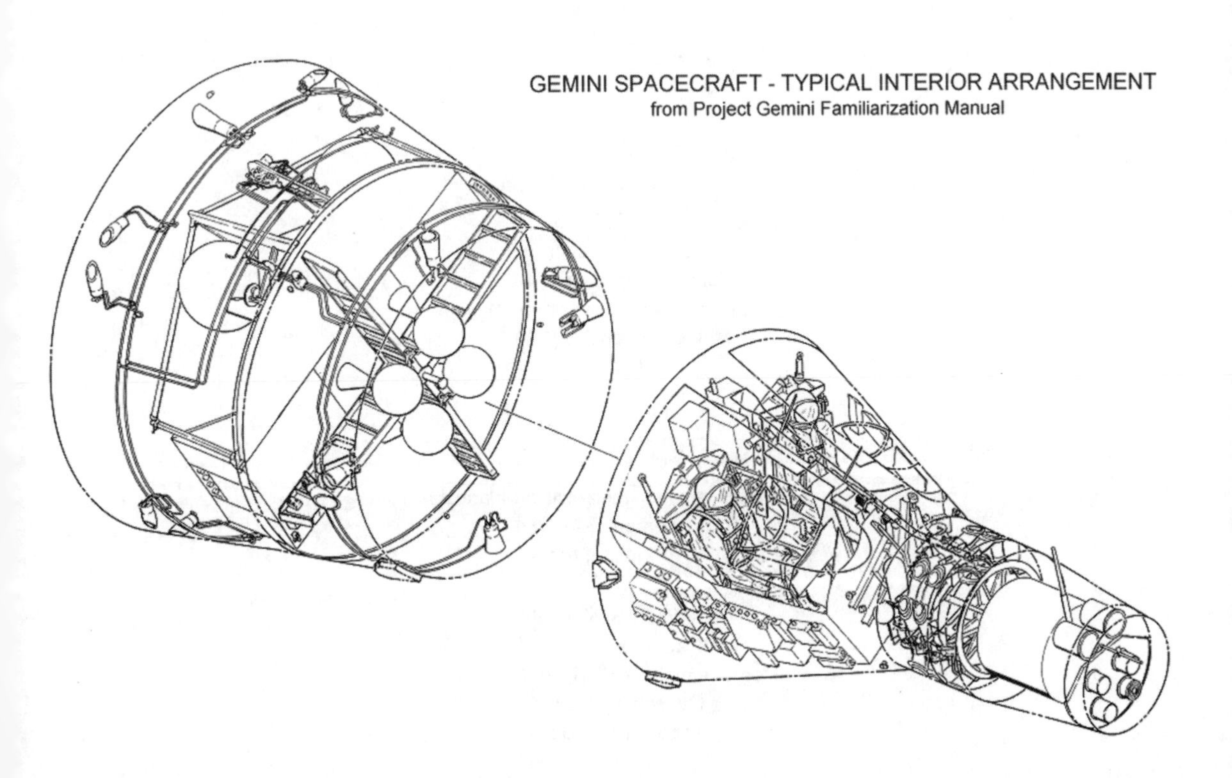

GEMINI SPACECRAFT - TYPICAL INTERIOR ARRANGEMENT
from Project Gemini Familiarization Manual

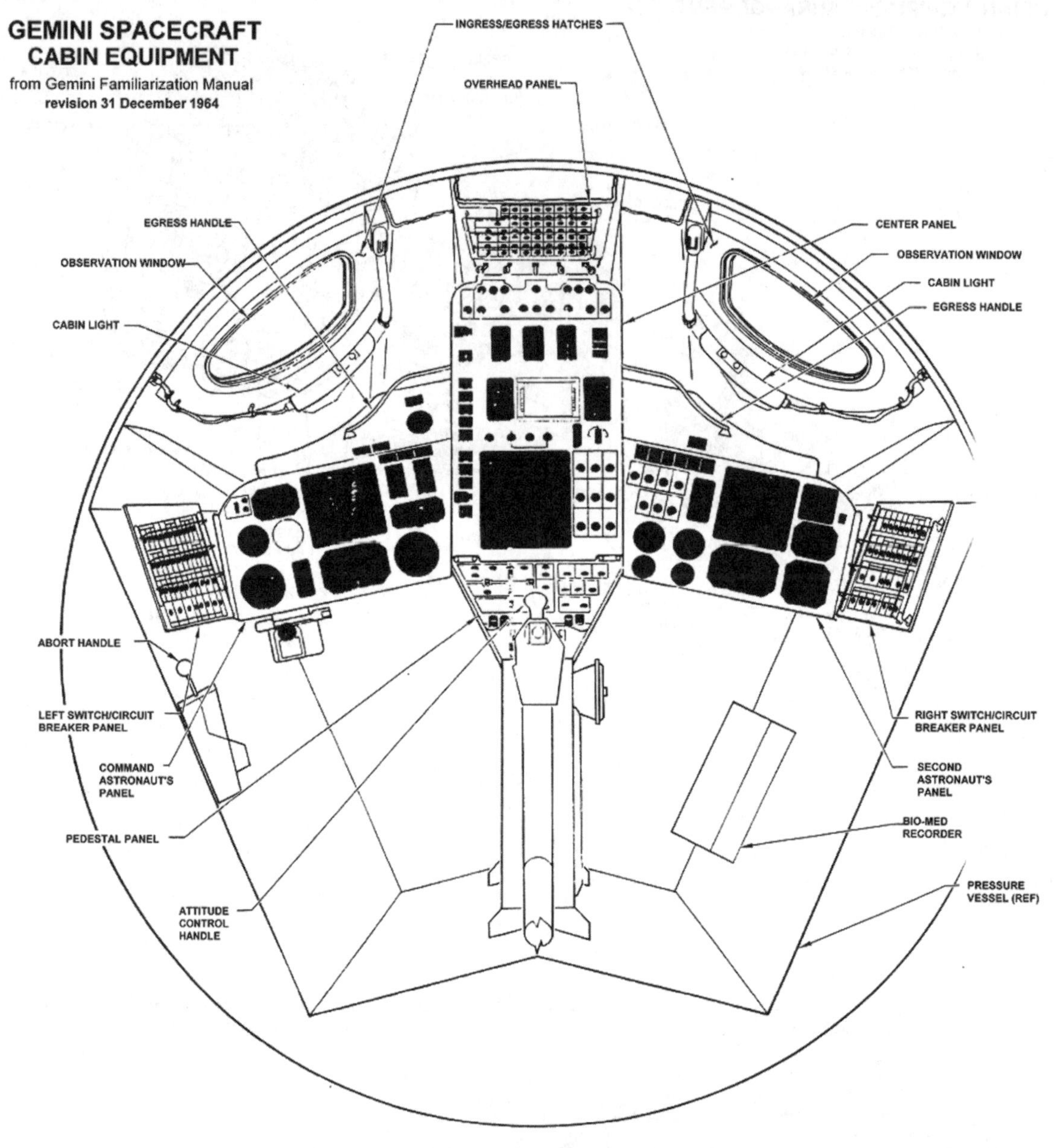

Gemini Spacecraft
Cabin Equipment (1 of 2)
from Gemini Familiarization Manual
December 31, 1964
(above)

Gemini Spacecraft
Cabin Equipment (2 of 2)
from Gemini Familiarization Manual
December 31, 1964
(opposite, top)

Gemini Spacecraft
Main Control Panel
from Gemini Familiarization Manual
December 31, 1964
(opposite, bottom)

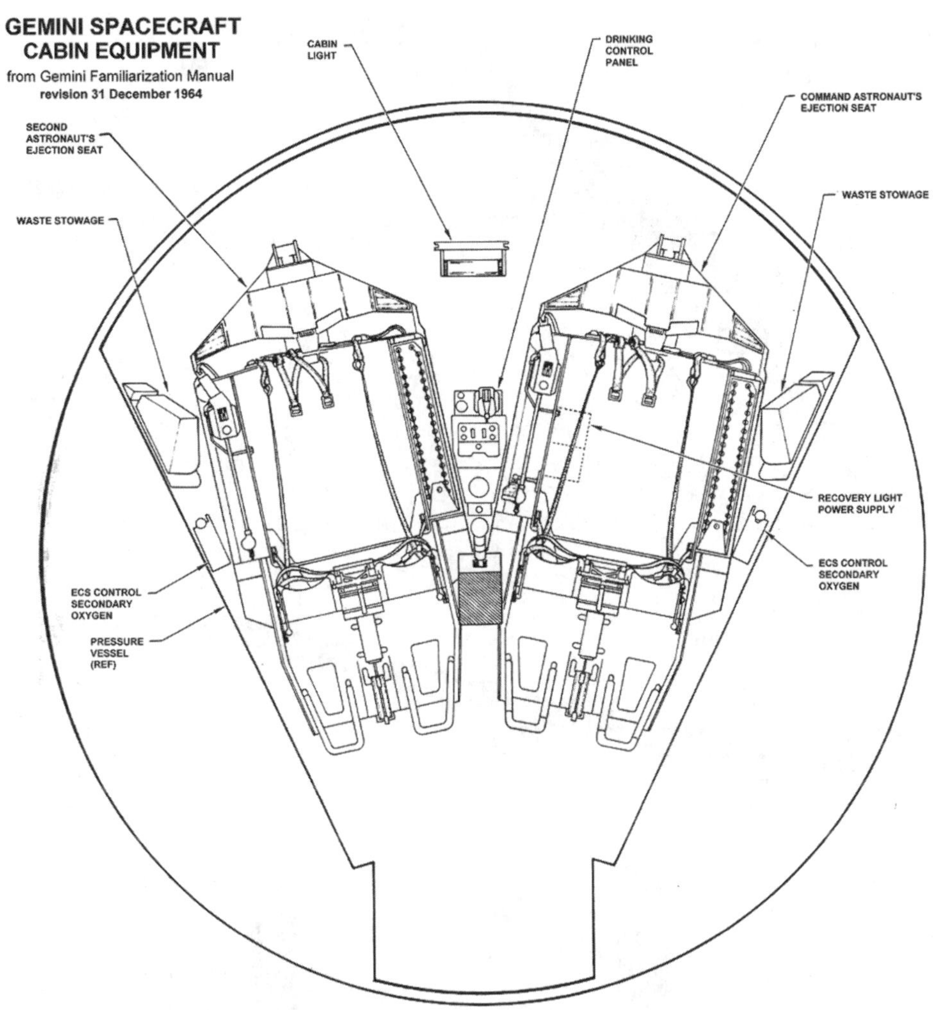

GEMINI SPACECRAFT CABIN EQUIPMENT
from Gemini Familiarization Manual
revision 31 December 1964

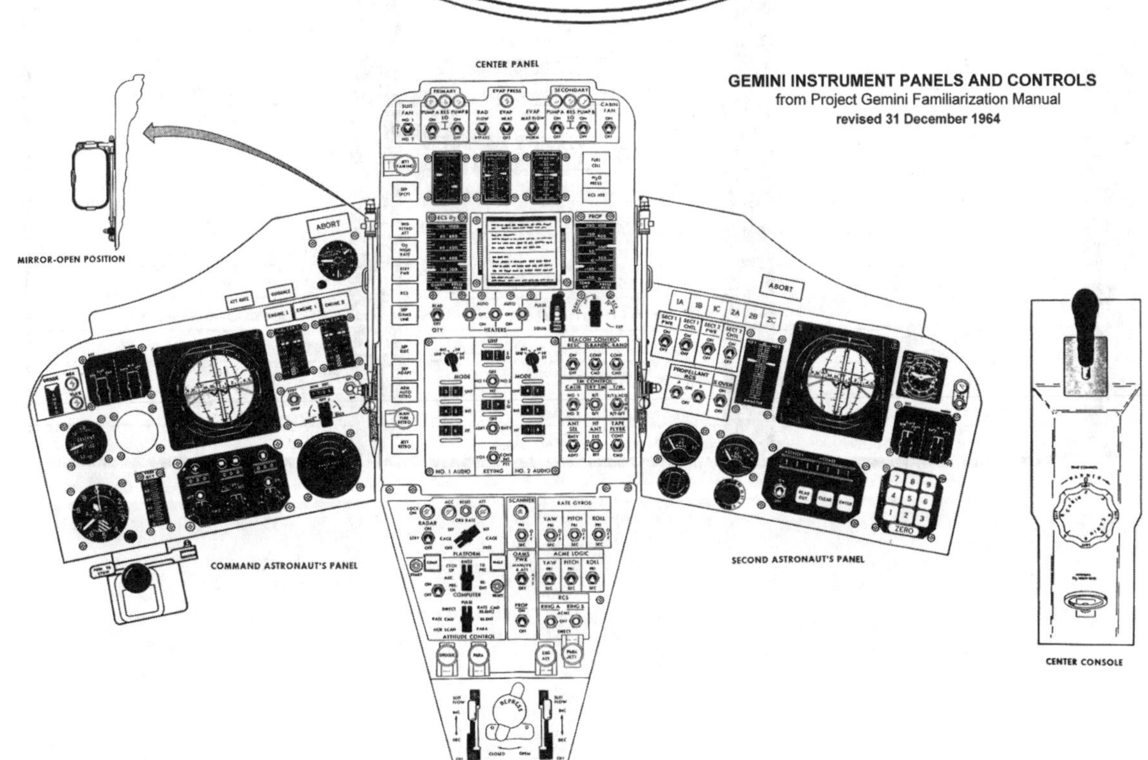

GEMINI INSTRUMENT PANELS AND CONTROLS
from Project Gemini Familiarization Manual
revised 31 December 1964

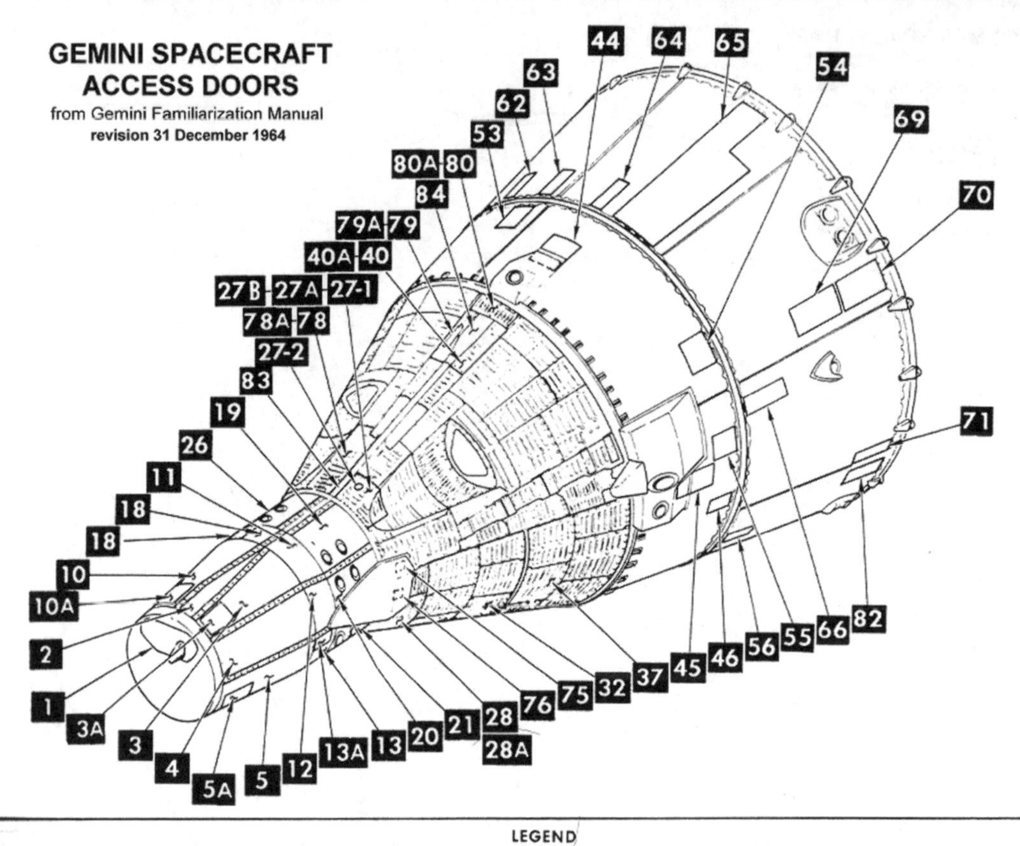

LEGEND					
NO.	DESCRIPTION	NO.	DESCRIPTION	NO.	DESCRIPTION
1	DROGUE CHUTE DOOR	26	RE-ENTRY CONTROL SYSTEM ACCESS	64	OAMS OXIDIZER PURGE ACCESS
2	DOCKING BAR CARTRIDGE ACCESS	27-1	SHINGLE 27-2 FRESH AIR DOOR	65	OAMS MODULE SERVICE ACCESS
3	SHINGLE	27A-27B	Z160.20 EQUIPMENT ACCESS	66	ECS SERVICE ACCESS
3A	DROGUE MORTAR CARTRIDGE ACCESS	28	SHINGLE	69	ECS PUMP MODULE SERVICE ACCESS
4	EMERGENCY DOCKING RELEASE CARTRIDGE AND GUILLOTINE CARTRIDGE ACCESS	28A	Z160.20 EQUIPMENT ACCESS	70	ECS PUMP MODULE SERVICE ACCESS
5	SHINGLE	32	FORWARD EQUIPMENT BAY DOOR-LEFT	71	SEPERATION SENSING SWITCH ACCESS
5A	RADAR ACCESS	37	AFT EQUIPMENT BAY DOOR-LEFT	75	ELECTRICAL DISCONNECT ACCESS
10	SHINGLE	40	SHINGLE	76	ELECTRICAL DISCONNECT ACCESS
10A	NOSE FAIRING RELEASE CARTRIDGE ACCESS	40A	RECOVERY LIGHT AND HOIST LOOP RIGGING AND CARTRIDGE ACCESS	78	SHINGLE
11	INTERFACE ACCESS	44	VERTICAL MANEUVERING ENGINE ACCESS	78A	Z160.20 EQUIPMENT ACCESS
12	INTERFACE ACCESS	45	LATERAL MANEUVERING ENGINE ACCESS	79	RECOVERY LIGHT DOOR
13	INTERFACE ACCESS	46	SEPARATION SENSING SWITCH ACCESS	79	RECOVERY LIGHT DOOR RELEASE MECHANISM
13A	GUILLOTINE CARTRIDGE ACCESS	53	OAMS LINE GUILLOTINE ACCESS	80	HOIST LOOP DOOR
18	INTERFACE ACCESS	54	F.L.S.C. TUBING CUTTER ACCESS	80A	HOIST LOOP DOOR RELEASE MECHANISM
18A	PYROTECHNIC SWITCH CARTRIDGE AND BRIDLE DISCONNECT CARTRIDGE ACCESS	55	FORWARD MANEUVERING ENGINE ACCESS	82	SHAPED CHARGE DETONATOR ACCESS
19	RE-ENTRY CONTROL SYSTEM ACCESS	56	FUEL CELL SERVICE ACCESS	83	COVER ASS'Y.-PARAGLIDER OR PARACHUTE CONTROL CABLES.
20	RE-ENTRY CONTROL SYSTEM ACCESS	62	OAMS OXIDIZER PURGE ACCESS	84	COVER ASS'Y.-PARAGLIDER OR PARACHUTE CONTROL CABLES.
21	RE-ENTRY CONTROL SYSTEM ACCESS	63	OAMS LINE GUILLOTINE ACCESS		

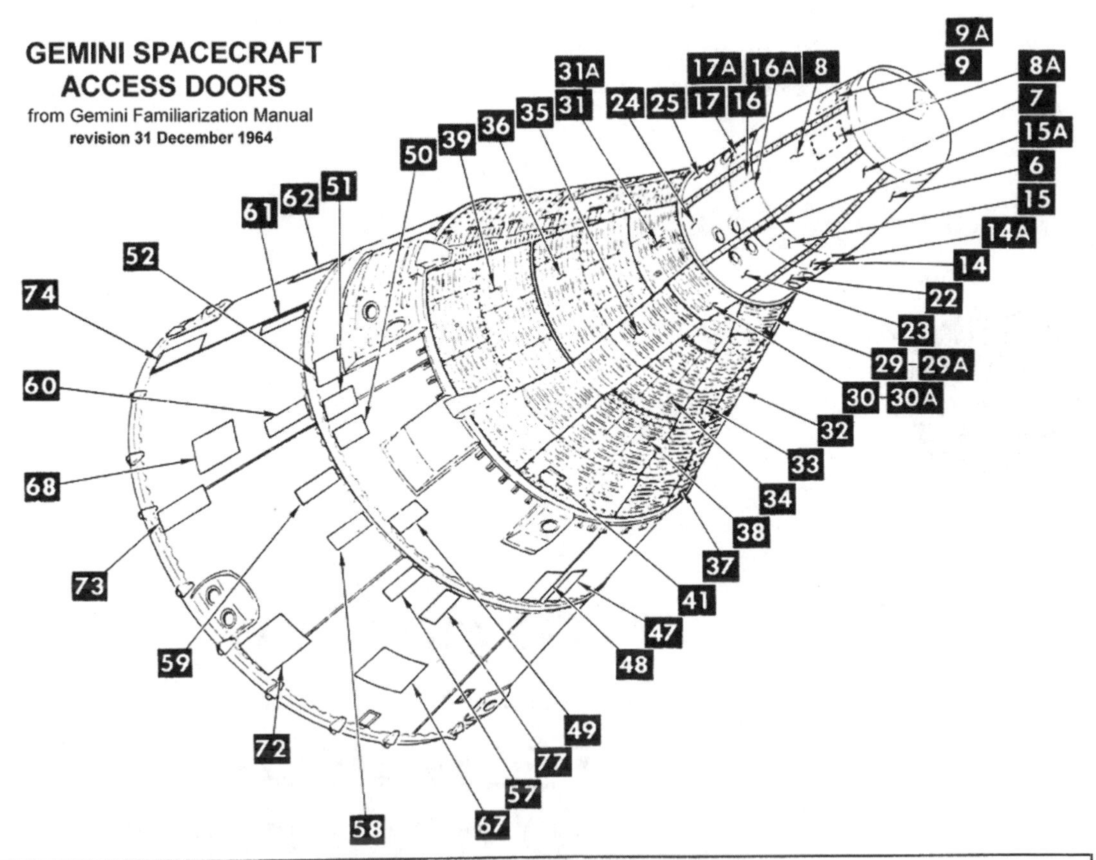

LEGEND					
NO.	DESCRIPTION	NO.	DESCRIPTION	NO.	DESCRIPTION
6	EMERGENCY DOCKING RELEASE CARTRIDGE AND GUILLOTINE CARTRIDGE ACCESS	25	RE-ENTRY CONTROL SYSTEM ACCESS	48	RELAY PANEL ACCESS
7	SHINGLE	29	SHINGLE	49	SEPARATION SENSING SWITCH ACCESS
8	SHINGLE	29A	Z160.20 EQUIPMENT ACCESS	50	GUILLOTINE CARTRIDGE ACCESS
8A	RADAR ACCESS	30	SHINGLE	51	B.I.A. RELAY PANEL ACCESS
9	EMERGENCY DOCKING RELEASE CARTRIDGE AND GUILLOTINE CARTRIDGE ACCESS	30A	Z160.20 EQUIPMENT ACCESS	52	FORWARD MANEUVERING ENGINE ACCESS
9A	DROGUE CHUTE DEPLOY SENSOR SWITCH ACCESS	31	SHINGLE	57	SHAPED CHARGE DETONATOR ACCESS
14	INTERFACE ACCESS	31A	Z160.20 EQUIPMENT ACCESS	58	FUEL CELL SERVICE ACCESS
14A	GUILLOTINE ANVIL ACCESS	32	FORWARD EQUIPMENT BAY DOOR - LEFT	59	GUILLOTINE CARTRIDGE ACCESS
15	INTERFACE ACCESS	33	MAIN LANDING GEAR DOOR - LEFT	60	GUILLOTINE CARTRIDGE ACCESS
15A	GUILLOTINE ANVIL ACCESS	34	CENTER EQUIPMENT BAY DOOR - FORWARD	61	OAMS FUEL PURGE ACCESS
16	INTERFACE ACCESS	35	MAIN LANDING GEAR DOOR - RIGHT	67	ENGINE TO SCUPPER INTERFACE ACCESS
16A	GUILLOTINE CARTRIDGE ACCESS	36	FORWARD EQUIPMENT BAY DOOR - RIGHT	68	ELECTRONIC MODULE TEST ACCESS
17	INTERFACE ACCESS	37	AFT. EQUIPMENT BAY DOOR - LEFT	72	GUILLOTINE CARTRIDGE AND LAUNCH VEHICLE ELEC. CONN. ACCESS
17A	PARAGLIDER ELECT. CONTROL BOX ACCESS	38	E.C.S. BAY DOOR	73	SEPARATION SENSING SWITCH ACCESS
22	RE-ENTRY CONTROL SYSTEM ACCESS	39	AFT. EQUIPMENT BAY DOOR - RIGHT	74	SHAPED CHARGE DETONATOR ACCESS
23	RE-ENTRY CONTROL SYSTEM ACCESS	41	PURGE FITTING ACCESS	77	FUEL CELL PURGE ACCESS
24	RE-ENTRY CONTROL SYSTEM ACCESS	47	RELAY PANEL ACCESS		

NATIONAL AERONAUTICS AND SPACE ADMINISTRATION

GEMINI VII

TECHNICAL CREW DEBRIEFING

December 23, 1965

Manned Spacecraft Center
Houston, Texas

PREFACE

This preliminary transcript was made from voice tape recordings of the Gemini 7 flight crew debriefing conducted December 19 through December 21, 1965 at the Crew Quarters, Cape Kennedy, Florida.

Although all the material contained in this transcript has been rough edited, the urgent need for the preliminary transcript by mission analysis personnel precluded a final edit prior to its publication.

1.0 COUNTDOWN

1.1 CREW INSERTION

BORMAN — I have no comment. I thought it went very well.

LOVELL — Likewise, no comment on crew insertion. I think we got quite a bit done. It was very orderly.

BORMAN — Timing was good and it was done properly.

1.2 COMMUNICATIONS

BORMAN — Communications were good. I had no trouble at all.

LOVELL — I had no trouble with communications in the cockpit or the spacecraft, but the communications in the van from the suiting area to the other area are rather poor. Maybe we should try to get that improved sometime.

1.3 CREW PARTICIPATION IN COUNTDOWN

BORMAN — Again, I think they have been used on 4 or 5 launches, and I thought they were fine.

LOVELL — Right. Countdown procedure and crew participation is just what you expect now.

1.4 COMFORT

BORMAN — Comfort was fine. No problems?

LOVELL — No problems for comfort, but I was surprised when I got in the cockpit, because there was a lot more there than there was when I got in it for the stowage review. But, it all turned out for the best. No problems.

1.5 ECS

BORMAN — ECS worked fine. We had no trouble with ECS at all during prelaunch or launch.

LOVELL — That is true. The purge was a lot slower and it was just perfect for the final countdown. It was too fast for the SIM Flight, which we went through, and I got an ear blockage. For the countdown, it was just right. Very slow.

1.6 SOUNDS

BORMAN — We had been well briefed on all the sounds: the gimbaling, pre-valve, and erector. As a matter of fact, when the erector started down there was no sound. We had been told that probably there might be a clanking or something. I heard nothing.

LOVELL — All I saw was the sky.

1.7 VIBRATIONS

BORMAN — Vibrations. No comment. I had no problems.

LOVELL — Is this liftoff vibrations?

BORMAN — No, this is countdown. Vibrations of the spacecraft during countdown.

LOVELL — No, nothing we had not heard before.

1.8
VISUAL

LOVELL — The windows were perfect. We had no fogging.

BORMAN — No fogging.

LOVELL — The windows were heated previously as a result of 5's problems, and our windows, I thought, were perfectly clean. Didn't you?

BORMAN — Right.

1.9
CREW STATION CONTROLS AND DISPLAYS

LOVELL — No comment. Exactly how we had planned it for months.

BORMAN — Exactly the way that we had seen it, and no problems.

2.0
POWERED FLIGHT

2.1
LIFTOFF CUES

BORMAN — Stoney came in loud and clear, counting the countdown.

LOVELL — Came in loud and clear.

BORMAN — We knew exactly when it was, and I for one had absolutely no question in my mind when we lifted off. It felt like I had been tied back, and someone cut the string and there was a slow but definite acceleration at lift-off.

LOVELL — I thought you could just about put CAP COMM, Vibration, and noise together, because the motion, vibration, and noise all contributed to a definite knowledge that you were going someplace.

BORMAN — In other words, what you are saying is that you had no problems determining lift-off.

LOVELL — No, it went.

BORMAN — Okay, vibration was nominal during lift-off. Again, perhaps it is because we were so well briefed on the simulations we have run, but I had no problems.

LOVELL — There was a little more noise than I expected, but a little less vibration.

BORMAN — Jim said there was a little more noise than he expected. Even so, it was not oppressive, or a problem at all. Visual. I did not have any visual cues. I was watching the instruments. What about you, Jim. Did you pick up any?

LOVELL — I had the clouds, and there was a visual cue. Just normal cloud cues.

BORMAN — Cockpit displays were good. The fuel pressure and oxidizer pressure were nominal the whole flight. Just perfect.

2.2
ROLL PROGRAM

BORMAN — The roll program, was so short it was almost like a spike. We hardly even noticed the roll program. Did you Jim?

LOVELL — I did not notice it at all. I heard you call it out, but I did not notice it.

BORMAN — I called it out, but we only rolled, I think, about 2 or 3°.

2.3
PITCH PROGRAM

BORMAN — The pitch program started just as in the simulator, which is very accurate on this. It looked exactly the same on the ball, and there was no problem.

LOVELL — Pitch program for the RGS followed exactly what the IGS was giving for the entire launch. The needles were just matched perfectly – nulled. I did not see any unusual attitudes that some of the other people commented on.

65-HC-1060

2.4
AERODYNAMIC

BORMAN — Again, we had had this described to us many times, and it seemed to follow right along. In the maximum q region we got some vibration and noise, but after we got through maximum q it was just like going supersonic in a fighter. You just slip through, and from then on it was just like riding on a train.

LOVELL — I don't think it was bad either.

2.5
ECS

LOVELL — Pressure went up to 5.5 in the initial stoppage, and it slowly leaked down to 5.1, and stayed there.

65-HC-1003

BORMAN — I was cool during lift-off.

LOVELL — I was too. Comfortable.

BORMAN — Comfortable. I mean cool in the sense that it means comfortable. Of course we can not very well comment on the cabin atmosphere because we were sealed.

2.6
MAXIMUM Q

BORMAN — We have already discussed this. There was some noise build up and some vibration, but nothing to worry at all about or even discuss.

65-HC-1061

2.7
WIND SHEAR

BORMAN — Wind shear. I did not notice any.

LOVELL — I did not notice any wind shear either.

BORMAN — You could not see any on you attitude gauges either, could you?

LOVELL — No, that is what I mean.

BORMAN — The attitude gauges stayed pegged. Right?

LOVELL — They stayed nulled throughout the entire flight. I was amazed at the accuracy with which the RGS was following the IGS program.

2.8
DCS UPDATES

LOVELL — Came through on schedule.

BORMAN — No problem?

LOVELL — No trouble.

BORMAN — Have any trouble punching the light?

LOVELL — After the second update, about 2:23, the g's are too high to let you punch off the light. So, you have to wait for staging, and then punch the light.

2.9
ENGINE I
OPERATION

BORMAN — Engine I operation, I thought, was normal. But I did notice a slight hint of a POGO around about, I would estimate, two minutes. The slightest, faintest hint. I do not think Jim even noticed it.

LOVELL — I did not notice any POGO.

2.10
ENGINE 2
STATUS

BORMAN — It seemed to me that from about 3 minutes and 30 seconds to around 4 minutes, the noise and the feel was a little bit different than it was after that, as if it was vibrating a little bit more. But this was sort of, again, a sensing type thing. The instruments were all nominal, and it may have just been me. I certainly can not complain about the operation.

2.11
ACCELERATION
G's

BORMAN — Any problems, Jim?

LOVELL — No problems. They were pretty nominal, weren't they? I could not see the g meter.

BORMAN — They were right on the money. And, of course, the g's we have are all experienced in the centrifuge, and so on. One thing, when the g's dropped at staging and at SECO I had no sensation of tumbling and no sensation of disorientation. Nothing at all.

2.12
POGO

BORMAN — I've mentioned that I detected a slight hint of one that was so small we cannot even really discuss it.

2.13
GUIDANCE
INITIATION

LOVELL — We had a guidance initiation. It was in the form of booster yaw deflecting downward, more so than booster pitch deflecting. Booster pitch deflected slightly to the right, indicating, at guidance initiation, a booster-high trajectory. But, they both came right back to null just after guidance initiation, and that was it.

BORMAN — We did not have the feeling that we were lofted, and then a sudden pitch down.

LOVELL — No, there was no change of booster performance at all. It was just that the needles deflected at guidance initiation to say that we had guidance initiation, and after that they nulled and stayed that way from there on.

2.14
BECO

BORMAN — At BECO, the whole spacecraft was engulfed in a red flame. I noticed that out of the corner of my eyes. Jim, you probably had a better view than I did.

LOVELL — Yes. Flames came up the side there to the window.

BORMAN — There was a definite, very brief instant of it, probably in the order of milliseconds, but it did envelope the spacecraft and I, in my own mind, wonder if this is not the place where we are picking up some of the smudge on the window.

2.15
STAGING

LOVELL — Well, I did not notice any smudge at the time of staging.

SLAYTON — You did not notice any?

LOVELL — I did not notice any. Of course, things were going pretty fast. I did notice it after we got into orbit, but not at that particular time.

2.16
ENGINE 2
IGNITION

BORMAN — Again, it is so well simulated that . . .

LOVELL — It is very smooth.

BORMAN — It is very smooth, and away you go.

2.17
RGS INITIATE

BORMAN — Well, we have talked about that.

LOVELL — Yes, that is what I was talking about back previously.

2.18
GO / NO GO

BORMAN — GO / NO GO. Houston, on the ground, came through great. We got a GO / NO GO before the 30 seconds we were waiting for spacecraft separate. So, we knew we were in good shape before we ever had the possibility we would have to burn. Of course, we also had the IVI's onboard and they are very good also.

S65-63766

2.19
SYSTEMS
STATUS

BORMAN — The systems were all great. No problem, during powered flight. We got two delta P lights.

LOVELL — Oh, yes, that is right.

BORMAN — We are talking about spacecraft systems. We got delta P light on BECO in the first stage that went off at staging, then came back on during second stage flight, and then the Section 2 delta P light did not go out and it was . . .

LOVELL — No, Section I went on and out again during the flight. It went out at, I think it went out at SECO.

BORMAN — That is right.

LOVELL — But Section 2 came on and we saw that one for the next 14 days.

2.20
ACCELERATION

BORMAN — Acceleration during stage 2 was right on the money, right on the program. I read off, I think it was about six and a half g's maximum. We read this off after SECO.

2.21
FAIRING
JETTISON

BORMAN — Fairing Jettison, I did not even hear it. I was concentrating on the horizon, trying to get set for turning around. Jim jettisoned the fairing and punched the Spacecraft SEP. I did not see anything or hear anything.

LOVELL — I saw debris and heard it and had a definite knowledge that a squib had gone off. There had been an explosion.

3.0 INSERTION

3.1 POST SECO

BORMAN — Maneuver controller was easy to reach. I had it out, and there was no problem. It came out and was ready to go. Attitudes and rates, there were none. The thing was as solid as a rock as far as I could determine. I was watching the horizon, and the attitude remained constant and the rates were so minimal you could not even pick them up. I noticed no transients, we experienced no . . . as far as I know that was discernible.

LOVELL — Did you try to damp out the . . .

BORMAN — There was nothing to damp out.

LOVELL — Okay.

BORMAN — In fact, I did not use the thrusters at all for that. It just sat there.

3.2 SECO PLUS 30 SECONDS

LOVELL — I have the IVI readings on a card. Do you have those cards that we took off?

BORMAN — Yes.

LOVELL — I am sorry. We did not get forward-aft, left-right, up or down because they were so quick, and I was trying to get the camera. But it was 17 in the fore and aft window, 13 in the left-right, and up and down was 20.

BORMAN — What do you mean you did not get them? They are there.

LOVELL — No, I did not know aft or forward, or left or right, or up or down.

BORMAN — Oh, I see.

LOVELL — I just saw that they were so small that I just wrote down the numbers as a . . .

BORMAN — 17, 30, and 20. There might have been a 13, 17, and 20 . . .

LOVELL — About what the numbers came up with.

BORMAN — Spacecraft separation. We separated with minimum delay between thrusting and Spacecraft SEP. Jim actuated the spacecraft separation. I did not hear the thrusters firing. I could not hear them; and I did not even hear spacecraft separation, but . . .

LOVELL — I heard Spacecraft SEP, but I could not hear the thrusters firing. But you told me you were firing the thrusters . . .

BORMAN — I said thrusting and SEP Spacecraft and we did it and away we went. I thrusted for about 2 seconds. Almost immediately, as soon as we had finished thrusting, I started a yaw right 180°, and the rates were right around, I think around — Of course, you should be able to pick this up off telemetry, but I would estimate they were 3° to 4° per second turning around. As soon as I had the booster in sight, I thrusted back 5 seconds. This is the way we tried in simulations. The simulations in St. Louis were excellent.

LOVELL — Turned out that was the best technique to use, 2 seconds for the 2 seconds forward and a 5 second return.

BORMAN — We turned around and there it was, bigger than the devil!

LOVELL — At that distance there was no problem staying in there.

BORMAN — Now, I did have some problem because the booster was bending so rapidly. It was tremendous. It looked like one of the autogenous lines had been cut. I guess it was cut with a pyro, and it was really bending and this was causing it to translate as well as rotate. And in order to stay with it, I was having to use quite a bit of fuel; although it was certainly a nominal task. I also went through several control modes switchings. I started out in PULSE and I could not get around fast enough, so I went to DIRECT and then slowed it up in RATE COMMAND. Slowed up the direct rate I was using with RATE COMMAND, and left it in RATE COMMAND without using the hand controller for a while. Finally went to PLATFORM. When I went to PLATFORM we had been off to one side of the booster. When I went to PLATFORM, it yawed me back around, and I lost sight of the booster. So we went out of PLATFORM and flew the rest of it in PULSE Mode using the reticule on the horizon for stabilization and using the maneuver controller for thrust. This is all on onboard tape, incidentally. The air to ground communications, throughout the flight were superior.

LOVELL — I was really amazed at the communications, especially the primary station. The UHF was outstanding.

BORMAN — We have already discussed GO / NO GO. They came through loud and clear before we ever SEP spacecraft. We had no need for a velocity correction.

LOVELL — As a matter of fact, right now would be a good time to mention that address 72 read . . .

BORMAN — Nominal was 25,804 and address 72 read 25,804.

LOVELL — Can you imagine that? Right to the foot! 25,804. I could not believe it when I punched it up.

BORMAN — The orbit quantities were given to us, I think, by Bermuda. Of course, at this time we really were not interested in them, although they were sort of nice information. We had a GO / NO GO.

LOVELL — It was 87-178 – the initial forward quantity that was called up to us.

BORMAN — The MDU readouts: Jim read 72 and when he saw it was 25,804, we had a GO / NO GO from the ground. I do not believe you even read the rest of them out, did you?

LOVELL — No, I did not bother reading out the rest of the addresses 94, 97, 52, or 73, because I saw the 72 nominal. I saw the IVI's were right in there so we did not bother reading out anything else.

BORMAN — Debris. I did not notice any debris.

LOVELL — I noticed debris. I was looking out at Spacecraft SEP and Jet Fairing, and noticed debris. I also noticed debris between the spacecraft and booster when we first turned around.

BORMAN — Could you identify any type of the debris?

LOVELL — No, pieces. That is all I could tell.

3.3 INSERTION ACTIVITIES

LOVELL — We followed the regular procedure.

BORMAN — We did not have any problem with safing our switches. No problem. I did not even stow my D-ring at insertion. I was too busy trying to stay on the booster, and I did not get it stowed for the first orbit, I guess, or half an orbit.

LOVELL — What we planned on doing was getting pictures of the D-ring. I got the bracket up at staging, and I actually had a minute after guidance initiate to reach back there and get the bracket and stick it up. It worked out very nicely before the g's started building up again on second stage. The bracket was up and in place and no problem at all. Then at SECO, I went around to pick up the camera, because we had the camera stowed where the Agena control box is located, I managed to get the camera up, and it was already plugged into the electrical wire. All I had to do was turn the auxiliary switch on, put it on the bracket, and push the button, and it started taking pictures. Just about that time, Frank mentioned he was going to start thrusting pretty soon so I had to go back and punch off the spacecraft. Then I read up address 72. So, I hope the pictures come out.

BORMAN — We were looking right into the sun; I hope they do too. The drogue pins were no problem. Jim got them, but again, not until well into the first orbit. As a matter of fact, I pulled my own yesterday morning there. The problem is solved; I think they are easy to get to.

LOVELL — They are easy to pull out.

BORMAN — I think that we have covered station keeping with stage II booster, partially. I will mention that the booster, being without attitude control, translating also with this impulse it was picking up from the venting, is definitely an order of magnitude more difficult than station keeping with a stable vehicle like Spacecraft 6.

LOVELL — First of all, you do not have anyone controlling the thing; you do not exactly know where it is going to go, and it might translate because it is venting and has a slight thrust.

BORMAN — I know a couple of times we got in a little too close and I backed out, because you just do not dare get as close as you do the way this thing is spewing. We got a real good picture, a good look at the nozzle. I thought that it looked like the nozzle was bent in on two places on the booster engine. It looked like the nozzle, the ablative skirt had been bent in. But then, it may have been just a shadow, because the next time I looked at it, it looked just like a new engine. The booster itself had no apparent damage. The only thing we could see was this big spewing where the venting was coming from. I did not see any venting from the roll nozzle at all. Did you?

LOVELL — No, the venting came from some line right along the bottom edge, near the engine section of the booster.

BORMAN — That is right.

LOVELL — It was a line of some sort that was open, and fuel was spewing out of it.

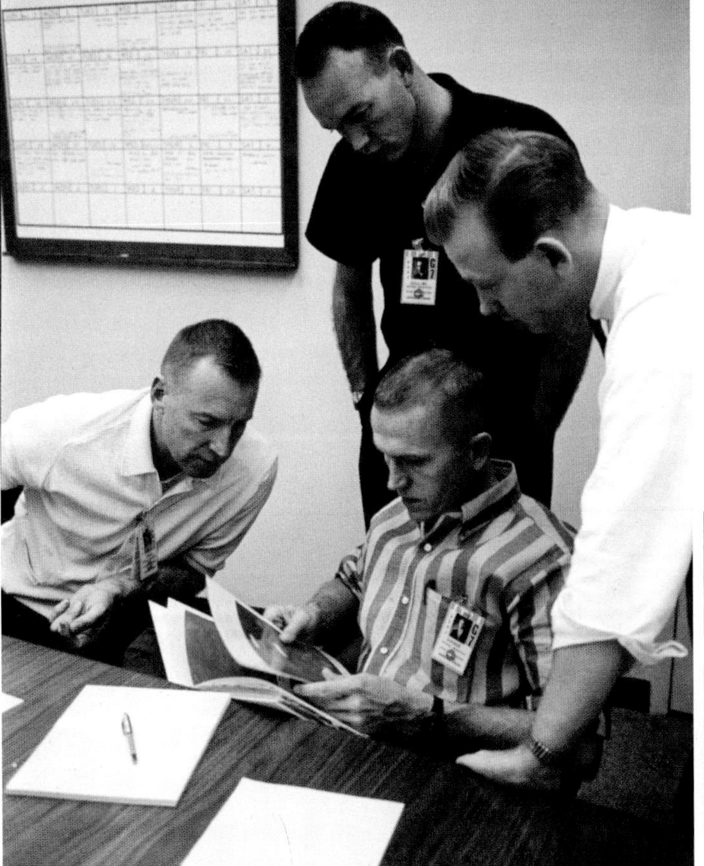

NASA Portrait of the Gemini 7 prime crew, Frank Borman and James Lovell (above). Lovell and Borman confer with the Gemini 7 backup crew Michael Collins and Ed White (standing, left and right, respectively). Fish-eye view of the interior of the Gemini 7 spacecraft (below).

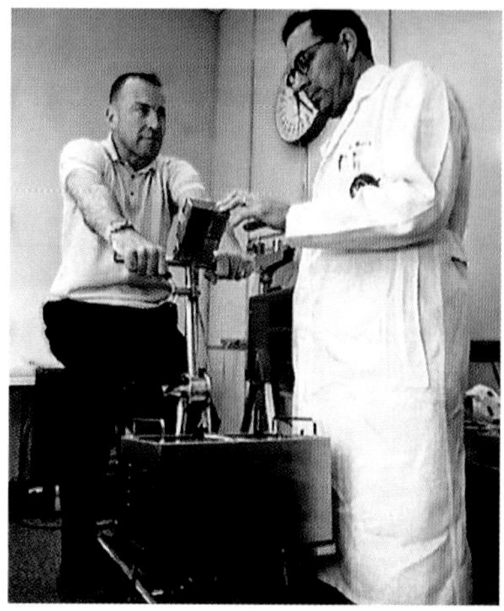

Astronaut Frank Borman during weight and balance test (left). Astronauts James Lovell during preflight physical (above). Fellow astronauts join the Gemini 7 crew for preflight breakfast (below).

Astronauts Lovell and Borman during water egress training (above). Astronaut James Lovell walks to the elevator on Pad 19 before Gemini 7 launch (below).

Launch of Gemini 7 on its Titan booster (above). Erector is lowered at Pad 19 during final minutes of Gemini 7 countdown (below). Astronaut Frank Borman in suiting trailer with Alan Shepard during prelaunch countdown (bottom).

Andes Mountains as seen from Gemini 7 spacecraft (top-left). Gemini 6 spacecraft as seen from Gemini 7 during station-keeping (left). Photograph of the Arabian Peninsula and the Gulf of Aden taken from Gemini 7 during orbit 130 (bottom-left). Gemini 7 prime crew during suiting up procedures at Launch Complex 16 (top-right). Technicians assist Gemini 7 prime crew in systems checks (above). Deke Slayton talks with newsmen after examining damage to Pad 19 (below-right).

Navy recovery helicopter preparing to lower the "horse collar" to the Gemini 7 recovery raft (above). Crewmen of the Gemini 7 spacecraft arrive aboard aircraft carrier (below).

Mission Control Center during Gemini 7 flight (top-left). Christopher Kraft monitors his console in Mission Control Center during flight (above). Astronaut Lovell talks to his family via radio-telephone aboard the aircraft carrier USS Wasp (below-left).

View of the Gemini 6 spacecraft as seen from the Gemini 7 spacecraft at approximately 38 feet apart (below). Astronaut Alan Bean in Blockhouse at Launch Complex 19 during Gemini 7 launch (bottom).

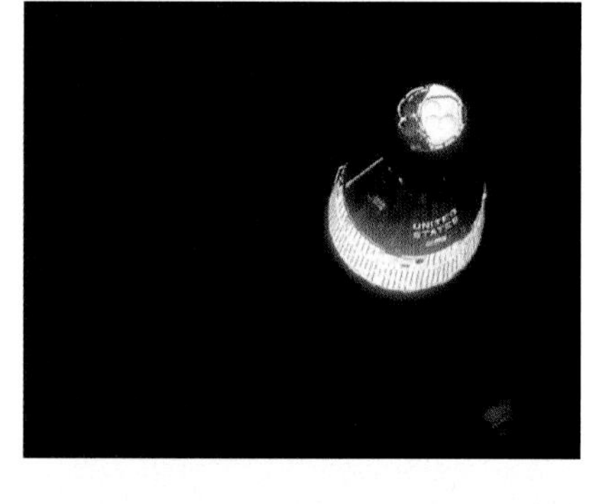

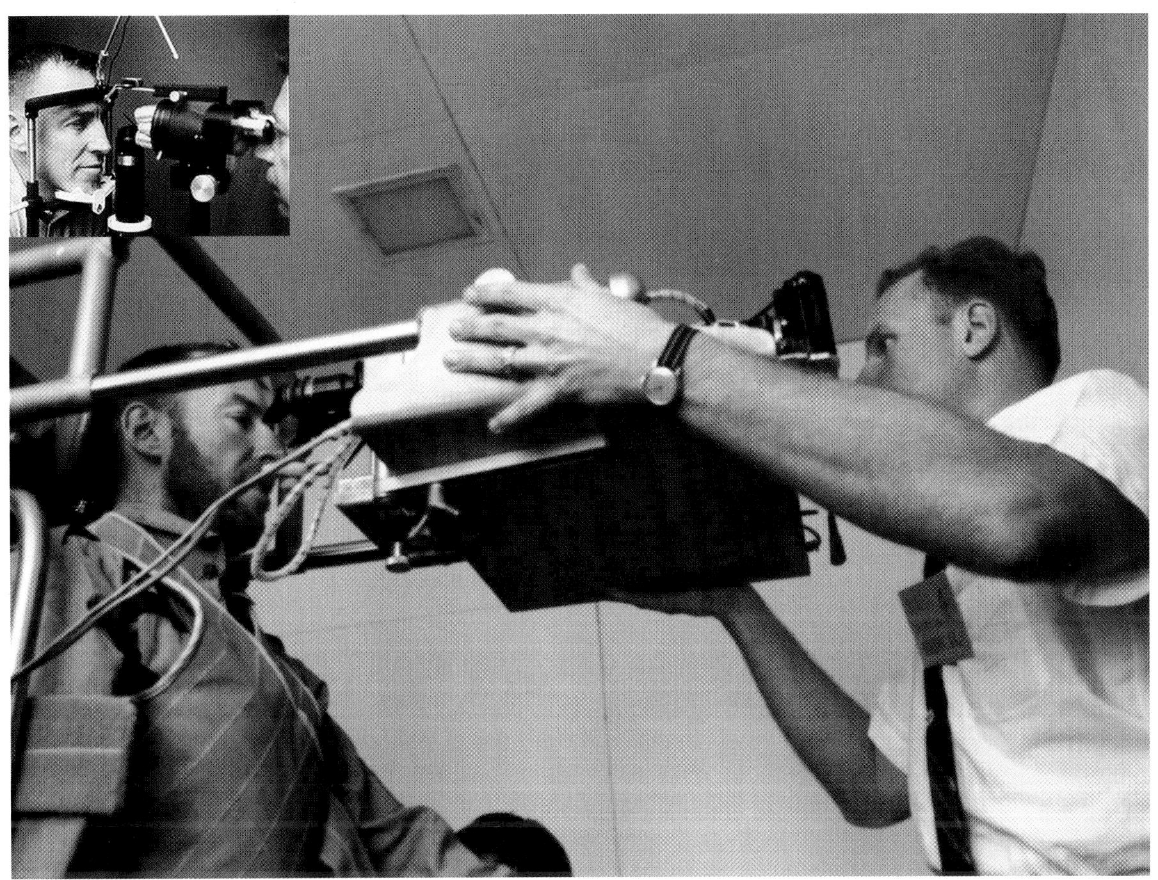

James Lovell undergoes vision testing postflight (above) and preflight (inset). Crewmen of the Gemini 7 mission during the welcoming ceremonies.

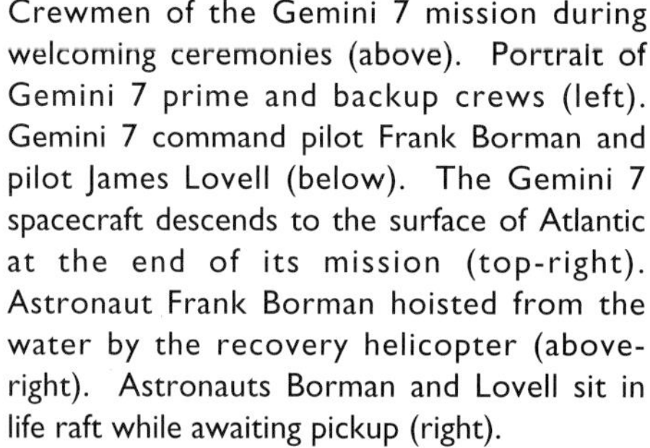

Crewmen of the Gemini 7 mission during welcoming ceremonies (above). Portrait of Gemini 7 prime and backup crews (left). Gemini 7 command pilot Frank Borman and pilot James Lovell (below). The Gemini 7 spacecraft descends to the surface of Atlantic at the end of its mission (top-right). Astronaut Frank Borman hoisted from the water by the recovery helicopter (above-right). Astronauts Borman and Lovell sit in life raft while awaiting pickup (right).

Crewmen of the U.S.S. Wasp watching the recovery of the Gemini 7 spacecraft (below). Borman and Lovell arriving with Deke Slayton and Alan Shepard for postflight debriefing (bottom-right).

BORMAN — I hope they got the data they wanted on the D-4 and D-7 Experiments. It was, again, a very uncomplicated maneuver, one that we practiced many times, and it worked just like. It does in simulation. Had no difficulty at all. The lights on the booster worked fine.

3.4
POST
STATION-
KEEPING

BORMAN — We did not do anything with stowage on the first orbit at all. D-ring, pins I have already mentioned, we did not get those in at all.

LOVELL — Arm restraints went down at 55 seconds. Belts. We did not even loosen them until after we had done D-4, D-7. The life vests we left right on the harness for the entire flight, but the harness did not stay on us for the entire flight. The sequence light test. This was done after the first orbit. We really had this insertion checklist in two phases, one at insertion and then one after D-4, D-7.

S65-63771

4.0
ORBITAL
FLIGHT

BORMAN — We have already discussed the station-keeping. That is no problem. I think the situation that we used, going off with about 2 seconds – 2 to 3 seconds – and thrusting back with 5 seconds while you are still on your side getting back to the booster as quickly as possible, solves the problem and takes a lot of the orbital mechanics out of the situation. I hope the film comes out. The one thing that did make it a little difficult on this one is when we looked back, we were looking back into the sun, and the booster was right in line with the sun. It was just like flying formation when the leader makes a turn, and you are down sun. It is difficult to see, and I tried to move off to one side and swing around and look a little bit more to the north. I think it was north. I guess I was trying to look to the south where I could get the sun out of my line of sight. I also had a cut-off on the booster at station-keeping at 88% fuel, so that at 88% fuel we were already in darkness, although we had not reached the time for the D-4, D-7 separation which was to occur at 00:25. I think it was about 00:23 or 00:21. So when we reached this limit and we were in darkness, I went ahead and separated, thrusting down.

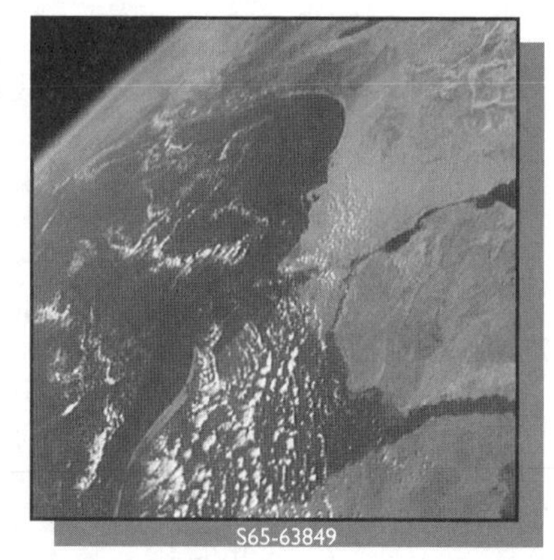
S65-63849

LOVELL — We actually separated earlier than 00:25. We actually separated at 00:21.

BORMAN — That is right. So we separated because we were in darkness and because we had reached the limit on fuel. We had been in darkness for a while. One thing I did notice was that the docking light was not particularly helpful on that stage of the business. I guess it is because we were not close enough to the booster.

LOVELL — We tried but the docking light just did not work.

BORMAN — I suppose because, again, we were looking at a lighted horizon with the docking light, and it did not work as well as it did later on with Spacecraft 6. The booster measurements went off. We got indications on the needle, on the measurement needle.

LOVELL — The recorder did not get on until 27 minutes. That is a guess. I am not too sure, but as I understand it, they had live transmissions up until that time, to Bermuda, and Antigua, wherever it is, so we were okay there.

BORMAN — The booster measurements were normal. Again, the simulator was perfect for that. The lights. Jim McDivitt had made some comment about not being able to judge distance because they only had two lights on there. We had four lights on and I will be darned if I will try to judge distance by four lights or fifty lights. You have got to have illumination or you have to have a stable vehicle.

LOVELL — You have to have something that illuminates the vehicle, not a light that flashes because you cannot tell from a flashing light.

BORMAN — Especially on vehicles rotating. I do not think that it is possible to control them. You have to have a controlled vehicle before you can judge distance from it, as far as I am concerned. The GO / NO GO, 17-1 T_R were no problem. We ran through the platform-off post station keeping checklist just the way it is listed.

LOVELL — Yes, that is where we caught most of the things.

BORMAN — That is where we caught most of the things like putting the D-ring away and the drogue pins and so on. Only one time in flight did we require attitude control fuel to change attitude for critical delay time playback. There was no problem. Communications, as always, were superior. The D-4 / D-7 Void Measurement was again no problem; just lined up on the black and ran for two minutes. Purging of the fuel cells. This is the first of a long . . .

LOVELL — Yes, but we did not do it then, did we? Did not we wait until we powered down and then waited two hours?

BORMAN — That is right. This is one of the things that they had in the flight plan that we asked them to change because . . .

LOVELL — Yes, we did not purge the fuel cells then.

BORMAN — Originally, this came right after power down and all of the fuel cell people recommended that you purge before power down, or wait until two hours after power down. So we did not do it at this time in the flight. This was changed. D-4, D-7 star measurements. There was no problem. The stars were well selected, and we were right on them. Right Jim? Jim copied down, on the procedures book, a check where we got the maximum return on the needle.

LOVELL — D-4, D-7 was a well organized experiment as far as Brentnall keeping us hopping about what to do. I will have to admit that.

BORMAN — He did a very good job. We knew just what to do. We had all the equipment with us and everything went very smoothly. MSC-2 and -3 turned out to be not much of a problem because at about the seventh day we turned it on and left it on for the rest of the flight. The Perigee Adjust Maneuver. Jim made the Perigee Adjust Maneuver. We did it on stars without a platform. I was timing for Jim and I think I fouled up. We planned to use a perigee to 102 miles, and I think we wound up with about 15 feet per second too much. It seemed like about 117 miles. One of the reasons that was causing this was we had come back into the vicinity of the booster, and just about midway through the burn the booster venting that was still occurring suddenly lit up, became lit up. It looked like we were flying through a lot of foreign objects or debris. I was afraid that we were going to hit something. At the same time this trailing wire came forward and slapped the spacecraft.

LOVELL — That is where I stopped.

BORMAN — Yes. After we had stopped and it hit us, I looked down and got confused and said, "No, we haven't burned enough." So we burned for about five seconds more. We had a trailing primer cord that would flop around and we didn't know what it was at the time, but it came forward when Jim stopped burning and flopped on the spacecraft. It made a noise and I thought we had hit some of the stuff that was spewing out of the booster. I wasn't sure that it was just fuel.

LOVELL — I think the ground people thought that this wire came forward because it had gotten in the way of the thruster fire. It definitely came forward after I stopped burning, because I stopped burning and this wire came slapping forward. It still had the momentum, you know. It slapped right in front of the window. I think the people got the impression that the thing had hit a thruster. It hit in front of us, then we stopped burning. But we stopped and then that thing hit and we added some more because we were still at apogee.

BORMAN — The first of many powerdowns was no problem. We went right by the check list. Some of these switch functions in the spacecraft, particularly toward the latter part of the flight, toward the 12th or 13th day – we were getting, I won't say lax in making them, but it seemed more of a chore to make these things right to the minute. Things like the BIOMED recorder and so on – we lost interest in having them turned off on the second. We knew what they needed to be turned on and off for. We didn't do as good a job from about the 10th day on as we did the first part as far as making those right to the minute.

LOVELL — As a matter of fact, why don't we get out the flight plan. I think we might have a lot of comments on it.

BORMAN — Let's start from the beginning.

LOVELL — The recorder was on at 27 minutes. D-4 / D-7 measurements. The GET of measurement that the COLD IR was outside the two degree field of view of the booster was at 30:13.

BORMAN — At 40:58 we had 84 % fuel left. We were right on the flight plan there.

LOVELL — There was another GET of measurement where the spacecraft was lined outside the field of view of the booster at 38:00.

BORMAN — We saw the booster for 2 or 3 revolutions after that. The lights were still working. We called it out and the ground got readings on this.

LOVELL — The Moon and booster were in view at 43:00. The booster and Moon were in view and we might get an erroneous reading because we were almost on the Moon.

BORMAN — Here we have a note that at 2:32 the fuel cell Delta P light blinked off at 2 hours and 30 minutes and then came back on. That is the section 2 delta P light.

LOVELL — Okay, then as far as stowage goes, the M-1 cuff was turned on at 3:03.

BORMAN — We put the bypass hoses on at this time also – the ECS bypass hoses. Incidentally, they turned out to be not too much of a problem. They were very handy for the type of work we did without suits on.

LOVELL — Right. We took the S/C out of the horizon at 2:08 to get some measurements, as requested from DOD, after we measured the stars. This is after we powered down the equipment. We connected the bypass hoses at 2:32. This was 2 hours plus 32 minutes.

LOVELL — Crew status reports. We had 3 or 4 a day.

BORMAN — 5:20, we started unpacking the meals. This is one thing that we had trouble with. Both left and right food boxes were jam packed. Fortunately, we changed the lanyards. We changed this during our stowage review, although it was difficult we got them out. Several of the meals had lost vacuum.

LOVELL — Which made them more difficult to get out.

BORMAN — Really you can't complain about this. The people did the best they could. We had an awful lot of food to store and we were able to get them out.

LOVELL — We had several blinkings of the Delta P light during this period. It went out at 6 hours, a little less than 6 hours, then came back on again at 6:27.

BORMAN — One thing that I wanted to find out about, and I still don't understand, is why we turned on the crossfeed valve right after launch. The FC O_2 pressure was just on the minimum of 150 psi at launch. I called up Houston and said I would like to leave the gauge in the FC O_2 position rather than the ECS O_2 position. Chris said, "No, unless we really felt strongly about it, they would rather have it in the ECS O_2 position." So we left it there and after we were inserted and we were still with the booster, they came in with a recommendation that we open the cross feed. When we did, this immediately raised the pressure to 250 psi. The thing that was bad was that we had over 100% oxygen and we were down to about 100 lbs. on the FC O_2. We agreed that we would fly at least 50 lbs. above the dome. So, I really didn't see the need for opening that valve although it didn't cause any problems.

LOVELL — They wanted to pump up and make sure.

BORMAN — It worked fine and we got right back up to 250 lbs.

LOVELL — That is one system that did work fine.

BORMAN — The first 7 hours was pretty nominal. All throughout the flight plan we have notes that the Delta P light went out and came back on and so on.

BORMAN — At 16:40 we sighted a satellite much lower and on a slightly higher inclination path than we were. It passed underneath us. It was so far away it looked like a sighting from the Earth. It was just a reflection. We were very religious about the exercise periods. We got those three times a day with the exception of the last day and one other day when we got only two. I think this is a very good idea. It is difficult and requires discipline because the last thing in your mind is the desire to exercise. You get lazy very easily. We did a very extensive operation with the bungee and also isometrics three times a day.

BORMAN — They were programmed 10 minutes. I think a more realistic one would be about five minutes, three times a day. I did 60 pulls on the bungee cord with both hands, 20 with each leg, and then ended up with 10 with each arm on the bungee cord in addition to the few for the crew status reports.

LOVELL — I did 60 pulls on the arms and 60 on each leg and it didn't make any difference. I could have done 20 on each leg and would have probably been better off.

BORMAN — At 45 hours Jim started taking off his suit. During that first 45 hours our noses were clogged and stuffy, our eyes were irritated, the cabin was hot; it was miserable. As soon as Jim started taking off his suit, the cabin even though he was out of the suit and I was in, got better than it was with both of us in our suits.

LOVELL — I didn't realize it was that long. We were almost up there two full days before I started taking the suit off.

BORMAN — At 49:53 we got a picture of Houston with the 250 mm lens. I hope it comes out. Okay at 69:40 we did a Perigee Adjust Maneuver, Delta V 12.4, 16.5 seconds, and came right on the money, using the stars, no platform. I don't think that there is any problem at all with the proper stars in making a gross adjustment.

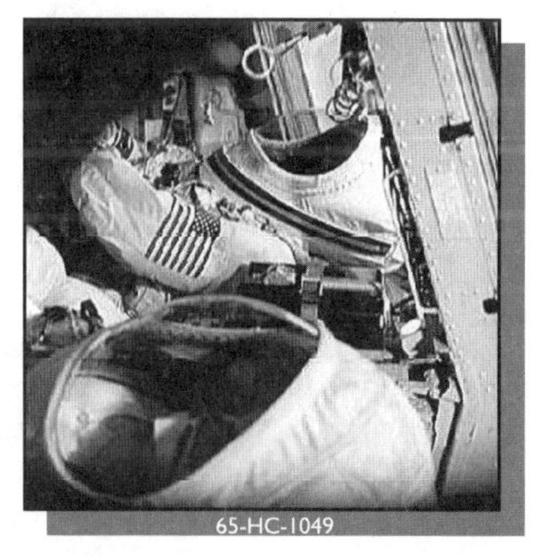
65-HC-1049

LOVELL — I think it was an excellent idea to do it without a platform, it takes two people. One person times and the other person burns on the star from attitude. Both people check the attitudes by looking at the star charts and getting the updates. Then making sure that the S/C is aligned right and the reticule is up to get the accuracy pretty good. After that, once you get it set in your mind what you are aiming at, one guy is in the cockpit with the watch or event timer and clocks it. The other guy has to look out the window because you can't go back and forth. If you look in the cockpit at the watch, you can't adjust to look out for the stars. So it takes two people for that. I think you can do a good job without a platform.

BORMAN — I do too.

BORMAN — There is one thing that was a pain in the neck, and I hope they get some good out of them, were UHF and the HF tests. That was an hour and a half transmitting every five minutes and having the HF / DF on. I'm not sure what kind of data they got but I hope they got something. The first one we had to do on the HORIZON SCAN; it took some fuel and I wonder really if it was worth it.

BORMAN — At about 166:40 we noted our drift rate picking up and we finally determined this was from the water boiler venting. It resulted in a left yaw rate and this continued periodically throughout the mission. It certainly would not be objectional if we had fuel to counteract it. During a night period, in which we didn't do any attitude control at all, I timed the rates during the 13th day, and when we woke up they were about 7 degrees per second. I timed them around the horizon and came up with 7 degrees per second. About the only thing you can say about it is that it requires fuel to stop it. It occurs primarily in left yaw and left roll.

LOVELL — There are two things in the S/C that causes the yaw left for some reason. Gus first noticed it and I think it is characteristic of the S/C. One is the water boiler and the other is, every time you turn off the power it fires two thrusters that give it a left yaw. The same two all the time.

BORMAN — We tried to beat that every way we could. Every time we shut down, we put it in a different control mode and it still fired the same two thrusters.

Every time you turn off ACME bias power it would go "boop," "boop," just like that. Every time we were without attitude control for extended periods we ended up with a left yaw and a left roll.

Finally at 191:48 we got both crewmen suitless. That was the best decision in the whole flight. The performance of the Cryo bottles was fantastic.

LOVELL — That was one thing we were worried about. The hydrogen bottle I thought was never going to last. Forty per cent of the hydrogen bottle was still left at the end of 14 days. One thing I wanted to try was to blow the squib. Remember they said "Did you blow the squib?" I forgot about it. Just prior to retro, I wanted to go over there and blow that squib that opened up into a vacuum.

BORMAN — It would have taken several hours for it to do any good.

LOVELL — Yes, I know, I just thought maybe we could hear it or something.

BORMAN — One thing that cropped up more and more as the mission progressed, it seemed to get worse as it went along was the fact that things were canceled because of weather. We picked up large areas of clouds over the U.S. and over South America. About the only area that stayed clear was Northwest Africa. A lot of the experiments and a lot of the Apollo landmarks were shot because of clouds.

BORMAN — On the 6 launch, the second time, we were able to track it. We were not able to pick up lift-off because of clouds again, but when it got to the con level, above the clouds, we were able to pick it up and we tracked it using IR until we couldn't see anymore. Even above the con level I think we were tracking the exhaust from the stage two engines using PULSE mode. I hope we got some good data on that.

BORMAN — At 266:16 we really got cold; the suit inlet temperature dropped below 40 degrees and we started squirting water out of the suit inlet hoses. We informed Houston about this and they determined that the water boiler had frozen up and they recommended a procedure to clear it. We did this with Gemini 6 watching; essentially it involved putting the radiator to BYPASS and changing some switches.

LOVELL — Evaporator heat on.

BORMAN — Put the evaporator heat on and setting up to 10 degree per second roll rate.

LOVELL — That's the picture you saw in the movies.

BORMAN — It actually threw a lot of fuel out and a lot of water out. It left a glob of ice on the side of the S/C, about 10 inches in diameter at the exit from the water boiler vent.

LOVELL — There were only two problems that we really had. There were the Fuel Cells and the two thrusters. We also had a cold Spacecraft.

BORMAN — Yes, that is when we had that water boiler problem.

LOVELL — Before that; the first time we woke up, it was 20 degrees colder inside.

BORMAN — Oh yes, I'm sure what had happened during the night was that we vented the water boiler, used the water boiler. This is the day when we woke up and had such high rates on the S/C. We have all that in the cabin temperature survey. The wall temperature was 20 degrees lower.

LOVELL — It was just freezing in there.

5.0 RETROFIRE

5.1 TR-2:00 POWER UP AND ALIGNMENT CHECKLIST

BORMAN — We had a slightly different procedure as far as retrofire goes. Powering up for it took two hours. The power up and alignment checklist was called up from the ground since we had open circuited two stacks. We turned our main batteries on and the squib batteries back on at TR minus two hours.

LOVELL — During the flight they had powered us down on the squib batteries and put in the bus ties about the last week of the flight.

BORMAN — We were flying with bus ties and fuel cells and no squib batteries.

LOVELL — To conserve the squib batteries for the retrofire period.

BORMAN — Right.

LOVELL — Because of that configuration, and because of the fact that we lost two stacks, we had to modify our power up procedure.

BORMAN — Right. Incidentally because of the fact that we had two degraded thrusters, 3 and 4, we didn't use the PLATFORM mode at all for this alignment. We aligned it all manually. The thrusters were degraded, but there was still enough in them to allow you to get fine maneuvers, fine control. I used less control by turning off the circuit breaker for thruster No. 12 and used 11, giving back thrust and this would give you right yaw.

5.2 TR-26 EVENTS

BORMAN — At T-26 the event timer was set, we didn't read anybody because of our orbit, and we weren't able to start our event timer counting down until T-20.

LOVELL — T-20.

BORMAN — Read out from Carnarvon.

5.3 TR-5 GMT STOP CLOCK

BORMAN — TR-5 Jim got the bug on the eight minutes, no problem.

LOVELL — Yes, I got that okay.

5.4
TR-256

BORMAN — TR-256 Sequence light came on exactly on schedule.

LOVELL — The digital clock never lost a second during the entire flight.

BORMAN — We didn't touch it.

LOVELL — We didn't touch that digital clock one time during the entire flight. That is the best instrument in the whole S/C, especially for this type of flight when you have a lot of updates and everything.

BORMAN — Electrical was no problem. Control system, the RCS worked perfectly. It just worked beautifully.

5.5
TR-1

BORMAN — Retro attitude minus 20 degrees pitch. The ball had been aligned for two revolutions and it was perfect. If we had not had the ball, I would have been happier if we had retro fired in the daylight. SEP OAMS, as advertised. You hear it.

LOVELL — Yes!

BORMAN — You feel it slightly.

LOVELL — That is right, and you even feel SEP ELECT.

BORMAN — Yes, and you really feel SEP ADAPT. It felt like I had put in forward thrust at that time.

LOVELL — Yes.

BORMAN — It was really a good thud when we separated the adapter. Retrorocket squibs were armed at TR-30. Arm AUTO-RETRO was actually done at about TR-10.

LOVELL — We did that a little bit early.

BORMAN — The event timer was perfect. MDU, Jim got all the readouts and they were exactly what had been called up.

LOVELL — There was one or two that the last digit was one number off, but that is nominal. We didn't bother that.

5.6
TR-0

BORMAN — From the time we got the countdown at Carnarvon we really didn't talk to anybody at all until we heard Houston at TR-10 seconds come in with a count through Canton.

LOVELL — We didn't think that they were going to come in, as a matter of fact.

BORMAN — No, we were wondering . . .

LOVELL — That is a very poor place to retrofire. Canton had poor communications compared with the rest of it.

BORMAN — But they came through that time.

LOVELL — Yes, they came through.

BORMAN — We really didn't need them because we had every indication that our timing was good on-board. They did come through but not until TR-10 seconds. At TR equals zero the S/C attitude was 20 degrees down. S/C rates were easy to control, but I thought that the thrust from those retro-rockets was high. I really had a sensation of being accelerated. Didn't you Jim?

LOVELL — Well, it was different from what I had expected because we were so used to zero g flight.

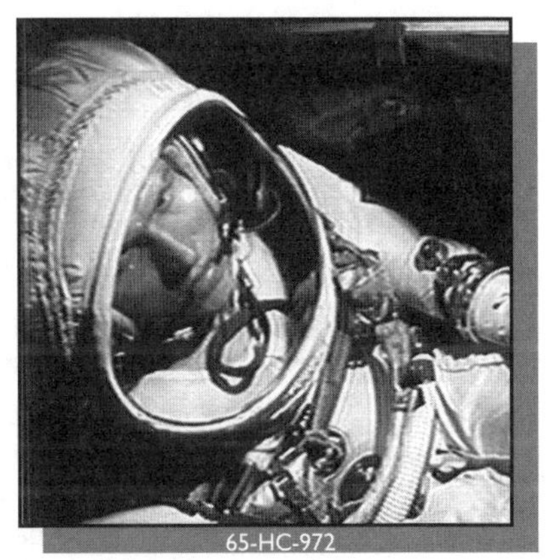
65-HC-972

BORMAN — The only thing I could do was fly instruments, the needles and the ball. Trying to hold it right on the ball.

BORMAN — I was very glad that I was in RATE COMMAND. I had to control it in RATE COMMAND a little bit, particularly on the fourth retro rocket. The first three went bing, bing, bing. Then there was a pause of about ½ a second and the fourth one went. The fourth one seemed like it was a little misaligned, I think it was left yaw. I had to bring it back. I would like to emphasize this. I thought those retros were really powerful, and that you were holding on to something that if you really didn't have good control it could get away from you pretty easily.

LOVELL — But, I was sure happy to hear them go.

BORMAN — Control mode was RATE COMMAND, and the IVI readouts there did you write those down?

LOVELL — I have them here.

65-HC-974

BORMAN — We called them off and we have them.

LOVELL — This is what I've been using. It was 298, and 112.

BORMAN — And 3 left.

LOVELL — Yes, and 3 left.

BORMAN — What were the nominals? Let's just make a note of what the nominals were.

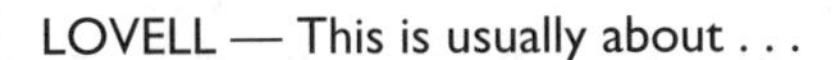

LOVELL — This is usually about . . .

BORMAN — They called up the nominals.

SHEPARD — They were 113 and 296.

LOVELL — Yes. That was 2 off from nominal, I recall that . . . 298 actual, and 112 actual, 298 aft and 112 down as the actuals.

BORMAN — And 3 right.

LOVELL — And 3 right.

BORMAN — So we got in close to the nominal, and when you figure this out on our onboard charts you come up with a bank angle of 50 degrees.

LOVELL — That's why I couldn't understand the 35 – well, maybe I'm wrong but let's take a look at this thing again. Let's go through it.

BORMAN — All right.

LOVELL — That's a minus 1 error here, right? And a plus 2 error there, right?

BORMAN — Right.

LOVELL — Okay, so I went in here and got to a plus 2 error here, right?

BORMAN — Right.

LOVELL — Went up here to a minus, here's the zero mark right here, to a minus one error; where this thing crossed this thing right up to here, plus 2, and by gosh, it came right out to 50 degrees or 53 degrees.

BORMAN — They gave us 55 degrees roll left, which is what the nominal level was . . .

LOVELL — I've got it right here. Fifty degrees and 60 degrees is what they gave us. Bank left 50 degrees and bank right 60 degrees.

BORMAN — Yes.

LOVELL — And so I looked up the chart and it said 50 degrees as the back up angles – everything was working just like a charm and then I went back here to the bank contour line to get out our down range deflection, and it was 1 or 2 miles, I think it was, no, 5 miles overshoot; which was just about as close as you can hack it. And I thought oh boy, this really, talk about nominal reentry, this is the one that's going to be it, and then they came up with 35 degrees, 45 degrees, and I misinterpreted it; I was arguing with Frank after retrofire and he says no, that's 30 degrees – 50 degrees.

BORMAN — 50 degrees.

LOVELL — 53 degrees is what he's saying. He just wants to get it down to a little finer line. And then Frank called back again and said, "No, it was 35 degrees," so I don't know what the story was there.

BORMAN — The FDI as far as the retrofire goes, it was no problem. It worked out fine, and I just like to have it, I think. If you really were forced into it you could do it on rate needles, but you'd have to have a lot of confidence in your ability to hold it. I wouldn't want to do it without RATE COMMAND; and again, I did it in RATE COMMAND. I'm not even sure how much the thrusters were firing during retrofire. Did you notice? I was watching the ball, and I didn't notice.

FCSD REP — Did it light up the horizon pretty badly?

LOVELL — It was really not too bad. But actually, yes, it did, it lit up quite a bit.

FCSD REP — Okay.

LOVELL — There was a point in the flight plan that they wanted the Pilot to evaluate the horizon for a night, no platform, retro. And the thing is this: you can turn out all the lights, you can get lined up for BEF retrofire, without a platform if you get the stars and everything. But once you start firing, you are going to have to use the rate needles, if they are working, to hold position, because you can't see the horizon any longer, because the thrusters do blank out any sight outside. And also, if you've got the lights turned up in the cockpit, so that you can see things; that means that you can't see outside. So, you have to go either outside to get cues, or you have to turn the lights out in the cockpit. And if you're going to use stuff inside, then you have the lights on. I would be hesitant to make a night retrofire without platform too. I think I would probably wait for a day one.

5.7
RETRO PACK
JETTISON

BORMAN — The retro pack Jettison – Jim fired . . . the one thing here on manual fire, Jim fired the manual retros the way we always have. We fired in the way we always have, one second after TR equals zero, but we got an auto retrofire.

LOVELL — Yeah, because the first one fired before I pushed the button.

BORMAN — That was right on the money. The retro pack jettison was accomplished 45 seconds after, when the amber light came on, and you could feel and hear this one going; of course, it was pitch black so we couldn't see a thing. This was one of the things that we didn't see, the RETRO ADAPTER, the ADAPTER, or the RETROPACK.

LOVELL — No, I didn't see any of that stuff go at all.

BORMAN — Total darkness.

LOVELL — Besides that, the thrusters blank out anything you could possibly see.

5.8
COMMUNICATIONS
AND
5.9
UPDATING

BORMAN — Communications were rather sketchy there. I was very glad though, that we were able to get through to Houston. I think it was over Guaymas when they came up and told us to change in retro angle, and bank angle; I don't know who did that but that was good work on the ground following up that computing, and getting us real time updates, I guess they must have done it after tracking.

LOVELL — Yes, that's probably what it was.

BORMAN — That's probably how they did it. And that was darn good.

LOVELL — Yes.

BORMAN — Because the 35 degrees, I was flying right between 35 degrees and 0 degrees most of the time, and if we'd have followed the 50 degrees, we'd have ended up way short. So that was very good work on the ground's part.

LOVELL — It looks like the initial computation of retrofire time was off, and they already had a good orbit on us.

BORMAN — I don't know what it was, but they corrected it when we came in.

5.10
POST RETRO
JETTISON
CHECKLIST

BORMAN — The post retro jettison checklist was accomplished with no problem. Oh, I'm thinking; we did have some discussion about as far as the retro goes. With the – we'll cover this more fully under suits. The question was whether to leave those hoods on or off for retrofire. We found that the noise and the – I don't know why we didn't notice this at launch, but we did during reentry, the noise from the air blowing in the G5C suits was an impediment to crew discussion.

LOVELL — It would go on the mikes and make a lot of noise on the mikes. The mikes picked up a lot of whistle.

BORMAN — Plus the fact that the vision out of that thing certainly needs to be improved. So, we didn't know what to do – we finally decided to leave them on for retrofire.

6.0
REENTRY

6.1
REENTER
PARAMETERS
UPDATE

BORMAN — Reentry. 400,000 feet, we had that time updated; and at 400,000 feet I rolled left 55 degrees, because this . . . or fifty degrees, the value of the backup angles at that time.

6.2
400K

BORMAN — Spacecraft attitude at 400,000 feet was difficult to determine. We didn't have a horizon until we were below 350,000 feet, and I was having a lot of trouble trying to find it. Jim, you got the horizon first on your side.

LOVELL — Yes, the horizon came up first on my side. Well, we did not have it right at 350,000 feet, but we could look out between RCS firings and see the air glow, if you'd stuck your face right up there and look out. But when you're doing the reentry on the instruments you have the lights up so, one guy can't do it, you have to have two guys; one to look out and find out where the horizon is and . . .

BORMAN — That was a heck of a thing. I'd like to be able to cross check between the balls and the horizon once in a while to make sure that I knew exactly where we were. As it turned out this was a completely instrument reentry. We finally found the horizon and Jim would tell me yes, it's about in the right place. But I just watched the ball. And I think that it would be very difficult to back up a reentry by watching out the window. One person could provide backup guidance for you, and tell you where you are and what the bank angle looks like with the horizon. But, I don't think that a person that is flying the reentry can cross check between the ball and the horizon. I think you have to make your choice and live with it. Okay, roll commands were just like the simulator; time correlation was good. The guidance initiate came right on the money, and the needles jumped indicating an undershoot, a slight undershoot. From then on we just flew it the way we'd flown them a hundred times in St. Louis and in the simulator. I think we were very well prepared for this. I tried to fly it so that we took it downrange, and we got a slight overshoot indication on downrange of about 1 needle width, 1 dot. Then as we got down to around 2 g's or 2½ g's, I tried to start zeroing it out, so then when 3 g's came; the downrange was pegged right on the money. And we were indicating zero on the cross range. And at 3 g's I switched to flying the roll bug, and just zeroing the roll bug; and as we came on down further and further the downrange stayed zero, but the cross range started going off full scale. Well, this really doesn't mean anything because all the cross range is indicating is your percent of miss versus percent of capability. And down on that range what it was really doing was, we were coming in a little bit short, and it was sacrificing the cross range in order to get the down range cleared up, because the cross range was very small anyway.

LOVELL — There was a bias in the down range needles between his ball and my ball, and I think, that fortunately, my ball was the one that was out. Because when he was right on.

BORMAN — You said we were overshooting all the time.

LOVELL — Yes. He was right on – I said you were overshooting, it was about a needle and a half width bias.

BORMAN — Okay. The initial indication of g's, I remember Jim called over and said, "how many g's are we on now." I said "less than one" and you said "get serious." I think you couldn't believe it. The first onset you feel like you have about a ton on you, but then as it builds up it never seems to get much worse. It's almost as if it were a step function. As soon as you get the g you really notice it, and then you don't notice it much more. And the maximum g's that we pulled during the reentry were 3.9.

LOVELL — Yes, that amazed me. I actually thought we did pull more g's.

BORMAN — 3.9 g's. So, it was a long extended time.

LOVELL — Yes.

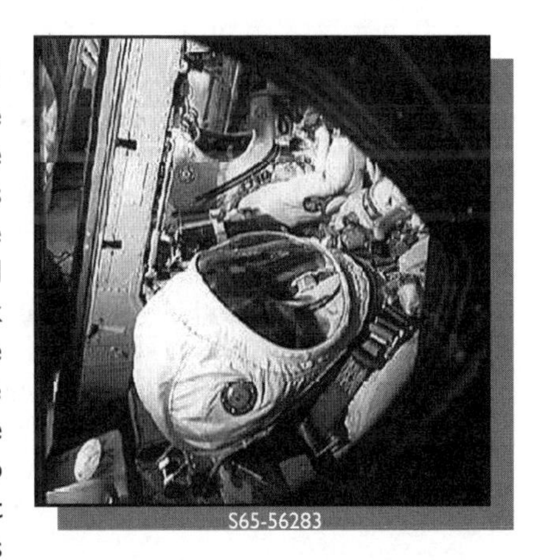
S65-56283

BORMAN — During the later part of it I started out in PULSE Mode and rolled over the 55 degrees in PULSE Mode, and then when we got Guidance Initiate I went to DIRECT. I was finding that in order to keep the cross range zeroed, and we had been told that Wally had trouble with his cross range, I was banking back and forth quite frequently maneuvering the spacecraft around the full lift point, from one side to the other and I was overshooting a little bit in DIRECT. I was also starting to pick up some pitch and yaw oscillations, so then I went to single ring RATE COMMAND. And boy, this was really a great control mode, it was steady as a rock. You could put it right where you wanted and it stayed there. But pretty soon we got down around, I guess it was when the g's were coming off, after 3.9 g's. I started losing it in single ring RATE COMMAND so I threw two rings on and it held it like a rock. But they were sure firing.

LOVELL — Oh, yes.

BORMAN — Boy, those thrusters were really firing. And we started getting ablation off the heat shield. It was coming back in and hitting the nose of the spacecraft, and that was pretty sensational. Jim was giving vivid descriptions on what was happening, and I was watching the ball.

LOVELL — That's one thing that no one had ever told us before. I was amazed. Maybe it was peculiar to the spacecraft.

BORMAN — No, Tom and Wally had mentioned it.

LOVELL — Oh, is that right? That ablative material went back and hit the forward end near the recovery section, rather the RCS section; and I thought well, I never heard of this before, and I was a little worried that maybe we were too far off, and the stuff was going to start getting into recovery sections. But it turned out to be okay.

BORMAN — Another thing was that the windows really got scrounged up on that reentry; that's something else. I could hardly see out of my window. Stuff was coming over from the heat shield and hitting it. It was really gunky.

6.3 ACCELERATION PROFILE

BORMAN — Okay. The acceleration profile was very small. It was a very high lift reentry, and or course, this results in a low g and long duration build up. No problem at all.

6.4 SPACECRAFT CONTROL

BORMAN — Spacecraft control was excellent until we got down to 100,000 feet or even below 100,000 feet. We turned on the LANDING SQUIB at 100,000 feet and sat there and watched it.

6.5 100K FEET

BORMAN — I started losing it; I think we may have run out of RCS fuel between 100,000 feet and 50,000 feet; at least I thought we had.

6.6 50K FEET

LOVELL — Well, didn't you turn off the RCS?

BORMAN — I didn't turn that off until after we got on the drogue. We were starting to build up the yaw and pitch rates. Then at 50,000 feet, I was anxiously awaiting the drogue, because these rates were building up a little. They weren't very bad yet, though. I pushed the drogue expecting it to decrease, and all it did was amplify them. And we got a real ride on the drogue for a while, sounded like the one Jim and Ed discussed. It was really going pretty bad.

LOVELL — Our angles were what? About 70 degrees. We pitched up?

BORMAN — Oh no, I estimate we were oscillating back and forth maybe 20 degrees.

LOVELL — From the drogue here pitching up we were rolling back and forth more than 20 degrees on that initial part.

BORMAN — We'll have to see. We don't have readings on that.

LOVELL — Okay, because I'm sure we did more than 20 degrees.

BORMAN — Then I threw the motor valves back open again on the thrusters, and that seemed to stop it. So I left them open a while and finally turned them on again and it stopped, and it settled out, and it was pretty smooth on the drogue. As a matter of fact, when we got down to main chute, it was steady as a rock on the drogue.

LOVELL — Yes.

6.7 35K CHECKLIST

BORMAN — I turned off the RCS motor valves and blipped the thrusters to clear the lines between 30,000 feet to 26,000 feet. And Jim then opened . . . The 40K barostat worked fine.

6.8 COMMUNICATIONS

BORMAN — And we gave the reentry status report. I'm not sure that Houston heard it, but we told them the drogue was all right and okay.

6.9
26K
CHECKLIST

BORMAN — Jim, at 26,000 feet you opened the vent air snorkel and we got a cabin full of I don't know what it was.

LOVELL — You had your hood off. Why?

BORMAN — I took my hood off to try to find the horizon, so I made the reentry with the hood off.

LOVELL — Okay, I had my hood on, and I think when we opened up the snorkel; the way that works, the snorkel draws air through the suit compressor, and then into the suit circuit, I had my hood on; and the flow comes out of an opening back here in the hood, and flows down. I got an eye full of something that was an acid.

BORMAN — Acid, eh?

LOVELL — Yes. Really burned my eyes. My eyes were watering when I finally got the hood off.

BORMAN — Well, we accomplished all the checklists, and we had no problems; as a matter of fact, it went pretty smooth in the time between the drogue deploy and the 10.6 barostat.

6.10
10.6K
BAROSTAT

BORMAN — It was just like the simulator. One thing I did notice, initially, when we were on the drogue, the altimeter was completely inaccurate. You couldn't even read it. We were oscillating so badly that it was jumping in thousands of feet per second, oh maybe not thousands, but the needles were going all over the place; and I remember thinking boy, if this oscillation doesn't stop, I'll have to punch the main chute on the amber light, rather than the altimeter. But the oscillations did stop.

6.11
MAIN CHUTE
DEPLOYMENT

BORMAN — I punched the main chute at 10,600 feet as indicated on the altimeter, and just a millisecond after that, the yellow light came on the 10.6 barostat light. The thing deployed immediately into a reefed condition, and we examined it in the reefed condition and it looked very good.

LOVELL — Frank thought it was in the reefed condition for 3 months.

BORMAN — It seemed like it stayed reefed for a long time, then it unreefed, and I couldn't find one gore or one panel that was ripped or frayed or anything.

LOVELL — It was a good chute.

BORMAN — Perfect chute.

6.12
POST MAIN
CHECKLIST

BORMAN — We accomplished the post main checklist, and then we braced ourselves very well and went to the single point of attitude.

6.13
SINGLE POINT
RELEASE

BORMAN — When we went to the single point attitude it was exactly the same as we had had it at St. Louis – where they'd rigged – they had a test after John and Gus's flight. They put a test capsule suspension at St. Louis, and this was exactly the same. You get a good whack and then you sit there and vibrate back and forth for a little bit.

6.14
2K
CHECKLIST

BORMAN — 2,000 foot checklist we accomplished with no problem. About this time, at 2,000 feet, I heard Air Boss calling and we started communicating with Air Boss. I saw him flying around while we were still on the chute. Houston came through about this time and wanted to know if we had a main chute. I'd called all these things off, but I guess that the communications – maybe the Auto Cats weren't working or something.

LOVELL — Air Boss should have called back and said . . .

BORMAN — But I called back and confirmed main chute.

6.15
LANDING

BORMAN — We hit the water with a pretty good thud, and your window went under water, didn't it? Jim's window went under water. We hit in a drift. We were drifting to the right rear, and there was a 14 knot wind, and when we hit the spacecraft rolled to the right, and your window went under the water, and mine stayed up. Nothing serious though.

LOVELL — Nothing serious.

6.16
POSTLANDING
CHECKLIST

BORMAN — We extended the HF antenna to get a test for them and went on HF-DF; I hope that somebody heard it. But they had swimmers there in about 4 minutes, and so I put the HF antenna back down to keep it from getting damaged. And we conducted the electrical check. I must say that I'm glad that the electrical check was simple, because it was hot in there, and we were tired. I was worried about this before and I would never have been able to sit there and go through this complicated, long check.

LOVELL — We had both planned, that what we were going to do was take off our suits in the spacecraft, and wear our orbital flight suits. And I think that we probably would never have gotten out of the suits, because we were just too hot and too beat.

BORMAN — It was even hot in the spacecraft, so Jim came through with the idea of opening the repress valve, and this was great. We had all that oxygen and you weren't going to use it. It blew all that cool oxygen out and we had O_2 HI RATE and the snorkel on. So we stayed pretty cool when you get right down to it. So it was a good idea. I don't know if you got the blood pressure measurements or not, did you?

6.17
BLOOD
PRESSURE
MEASUREMENT

LOVELL — I took them, I don't know whether they came out or not. I put the reprogrammer on in the water and started taking blood pressure measurements and – but that's hard to do. I ought to comment on that. Because to take a blood pressure measurement you had to pump up the thing and leave your hands still, and leave your arms still until it bleeds down; well it takes a little while for it to bleed down. Meanwhile, Frank's got the checklist out and the guys out there are putting on the collar, and we're trying to throw switches and take this and that; I just thought I might as well start doing it with my other hand.

BORMAN — Same way with the blood pressure they requested over Guaymas during reentry, I make a complete testimonial here; I think once the reentry starts that everything else gets left aside, and you don't mess around with blood pressures, or experiments or anything else. From then on it's sort of a case of surviving the darn thing. I didn't want him messing around looking for a blood pressure; so we didn't do it. About that time we couldn't find the horizon anyway.

LOVELL — We got called up from the MCC.

BORMAN — But, anyway, we didn't do it. So it didn't bother us.

LOVELL — That's the first I'd heard of it when they called up.

7.0 LANDING AND RECOVERY

7.1 IMPACT

BORMAN — We were drifting backwards, blunt end forward, rather, as we hit the water. Although it was a good jolt, I wouldn't say it was anything outstanding. We hit, and Jim, your window went under water, right?

LOVELL — The spacecraft rolled to the right, I believe.

BORMAN — Yes. We hit, rolled to the right, and you went under water and bobbed right up.

65-HC-1168

LOVELL — Right

BORMAN — I released the parachute and it floated in front of us for several minutes.

LOVELL — I saw part of it on the left hand side there, or rather on the right hand side as it floated by my window.

BORMAN — It stayed there for several minutes. It's just the way it's been described before.

LOVELL — There was an awful lot of fog on the window, though. I noticed that the humidity was such that you could hardly see out. Very foggy.

BORMAN — I'm not sure that was humidity or that was from reentry.

LOVELL — Might have been from reentry, I don't know.

BORMAN — I did see the S2F on the chute. We saw it while we were still on the chute coming down.

7.2 CHECKLISTS

LOVELL — The only thing I had about the checklists; during the recovery phase, I had a hard time doing the checklist, in fact, I had to give it to you, because I couldn't move my arm doing the blood pressure work. And that complicates the recovery phase of it quite a bit.

65-HC-1167

BORMAN — Yes.

LOVELL — I think it also compromised the blood pressures that way too.

BORMAN — We didn't feel like running foot races when we finally hit the water. We had planned to get out of our pressure suits into that orbital flight suit, but the effort was just too great. So we just opened the repress valve to get some more cooling in there and sat.

LOVELL — That's right. We opened up the repress valve; did we have the cabin fan on?

BORMAN — No, we didn't have the cabin fan on. We had the snorkel valve with the suit fans and the O_2 HI RATE and the repress valve open.

BORMAN — The checklists were all right then as far as you're concerned:

LOVELL — Yes. I thought the recovery phase was very good. I think I missed one or two. I know I didn't turn all the stack switches off, but the power and control switches were off during the reentry phase; so, there was no problem there.

7.3 COMMUNICATIONS

BORMAN — UHF. We had communications with Air Boss while we were still on the chute, and we had very good communications with them in the water. The communications with Houston via UHF were poor. Once we were on the drogue they kept calling us asking us to confirm main chute. I'm not sure they ever heard us confirm main chute.

LOVELL — I've often wondered about that. Watching the other spacecraft come in, why they don't call; and I found out that they do call but they can't get through. Must be the relay planes trouble or something.

BORMAN — Communications with recovery forces on UHF was excellent. HF: we extended the HF antenna, put out HF-DF tone for a while. Again I am not sure if anyone picked it up or not. We retracted it after it had been up only 8 minutes, because of the fact that we did not want to get the HF antenna broken off during the recovery operation.

LOVELL — There was no need for HF communications since we were so close to the recovery group.

BORMAN — The chopper was over us about 5 or 6 minutes after landing. We had much better UHF communications, so, we did not use HF.

On point of impact, onboard data. Within the limits of the readability of that scale, it was excellent. Down range and cross range needles were fine. We actually ended up about 8 or 9 miles from the carrier. You just can't get much finer information out of the down range and cross range.

LOVELL — Did you have any kind of a malfunction in the accelerometer?

BORMAN — No, but Spacecraft 6 did. I don't remember that being a condition of the bet.

LOVELL — I didn't either.

BORMAN — Ground Information. The ground gave us excellent information, as far as everything we needed to know, including recomputing the guidance angles after retro. The ground did an excellent job. Tracking data, I don't remember receiving that. When Spacecraft 6 was entering, they kept telling them that they were fine, and they were going right down the slot and everything. I do not remember ever hearing from the ground on anything like that on ours, do you? Perhaps we did and we were so engrossed in flying it, that we did not notice it.

LOVELL — Well, we had good communications prior to blackout over Guaymas. After we started guiding, going into the atmosphere, communications went to pot.

BORMAN — Status of recovery. I do not think recovery could have been any better.

LOVELL — Very smooth.

7.4 SYSTEMS CONFIGURATION

BORMAN — The ECS, as we said before, we had O_2 HI RATE with both suit fans, snorkel valve open, and the vent valve open. We also opened the repress valve. Electrical: We performed a simple electrical check. We turned off 3 and 4, left 1 and 2 on, and watched for the variation in voltage on the main buses. The bus that is fed by 1 and 2 batteries varied with wave action.

LOVELL — That is right. But 3 and 4 did not move from zero.

BORMAN — And then you turned off squib batteries 1 and 2 also didn't you?

LOVELL — I left squib battery 3 on.

BORMAN — Squib battery 3 was the only one that was on.

LOVELL — Control: We turned off the platform, the computer, the circuit breakers to the thrusters, and the RCS thrusters.

BORMAN — We left the computer in PRELAUNCH for 48 seconds or more, before we turned it off. Aeromedical, no comment. Except blood pressure being a nuisance, and perhaps even a hindrance when we were trying to go around the cockpit with the switches and you had to hold one arm still.

7.5 SPACECRAFT STATUS

BORMAN — RCS fumes: When you open the snorkel at around 26,000 feet you get a good load of them.

LOVELL — I am not sure what kind of fumes they were. They were not familiar to me. I have smelled the results of fuel in the RCS system, and I know what that smells like. It did not smell that way. I got a burning sensation in my eyes, which was different. Now, I might have got a more concentrated one. I still had my helmet on, and zipped up. You had your helmet off. I believe, that with the snorkel open, the compressor pulled the ambient air through the snorkel through the compressors and into the suit circuit. That is why I got a concentrated dose of whatever was on the outside, which caused my eyes to water and to burn. Whether it was the ablative material, the shingles, or the RCS fumes, I do not know.

BORMAN — The main chute was perfect.

LOVELL — Looked beautiful.

BORMAN — I could not see a rip or a tear or any fraying or anything; it was just perfect. The windows were foggy in flight. I thought they fogged over and the visibility out of them during the hot part of the reentry, was very poor also.

LOVELL — They started to burn a little bit. Started to peel off on the outside. I do not know what it was.

BORMAN — When we got on the water they were fogged over with humidity and salt spray. I guess you have to expect that. Leaks: There were none, that I know of. Couldn't see any or hear any.

LOVELL — I did not see any leaks.

BORMAN — Electrical Power: We mentioned we had 1 and 2 main batteries on, and when we evacuated the spacecraft, we turned all four of them on per the checklist. Turned off everything but the rescue beacon. Electrical power was ample, very good. We were running both suit fans.

LOVELL — Oxygen: Went to repress valve open.

BORMAN — We went through that swiftly, as a matter of fact, to keep cool.

LOVELL — We noticed before we got out that both the bottle pressures were down to zero.

BORMAN — After one of the swimmers said we were clear to open the hatches, I unlocked mine. It operated very freely and easily. I could budge it about 2 inches, but I could not lift it. I probably could have if I had exerted a lot of effort, and gotten my legs up under me. However, the swimmer was right outside, and I asked him to help. He helped and it came right open, worked very well. We had the suits on and left them on.

LOVELL — We were warm, undoubtedly. Getting out of the spacecraft as quickly as we did helped us.

BORMAN — That was the smart thing to do.

LOVELL — I wouldn't want to sit in there with our suits on.

BORMAN — Plus, I thought the visibility of that suit during reentry left a lot to be desired. That is why I had to pull my hood back to find out where we were and what position we were in. I think the suit is an excellent one, but it is going to have to be improved. We better grab it and start working, modifying it; to make it acceptable for Apollo. The sea condition was very good; 2 to 4 foot waves. We bobbed around, although I got a little queasy, I did not get nauseated, Jim didn't either.

LOVELL — The sea condition was outstanding for landing.

7.6 POSTLANDING ACTIVITY

BORMAN — Postlanding activity was well organized. We were a little busy. We did not get through until about 10 o'clock that night. Is that right?

LOVELL — Yes, that is right.

BORMAN — We had a little misunderstanding about riding a bicycle. We understood we were not supposed to ride until 18 hours after impact. They wanted us to ride it that night after we had been through a full day of medical exams, and finally had a good supper. So, we told them they would have to hold off until 18 hours after impact.

FCSD REP — You are still in the spacecraft for this part.

BORMAN — Okay. On postlanding we just sat there.

LOVELL — Well, we went through the check-off list. That took all the time. I saw the swimmers, checked the electrical system, that they wanted us to do for post landing. By that time the swimmer had the collar up. I could see the collar going up, and then he got the jacket on.

BORMAN — We had good communication with the swimmer through that jack.

LOVELL — Excellent communications with the swimmer.

7.7 COMFORT

BORMAN — It gets pretty warm in that spacecraft. I would hate to spend any great deal of time in there without any ECS.

LOVELL — I can speak as an authority on that.

7.8 RECOVERY FORCE PERSONNEL.

BORMAN — We covered communications, it was excellent. Flotation collar was fine, worked good.

7.9 EGRESS

BORMAN — Egress was normal, just as we practiced in Galveston Bay several times. These helicopters did a fine job. I think someone said it was about 23 minutes after we landed that we were on our way back.

7.10 SURVIVAL GEAR

BORMAN — Even the underarm life preservers inflated this time. Wonder of wonders. No problem.

7.11 CREW PICK UP

BORMAN — The crew pick up was nominal.

LOVELL — Nothing else. Everything was fine.

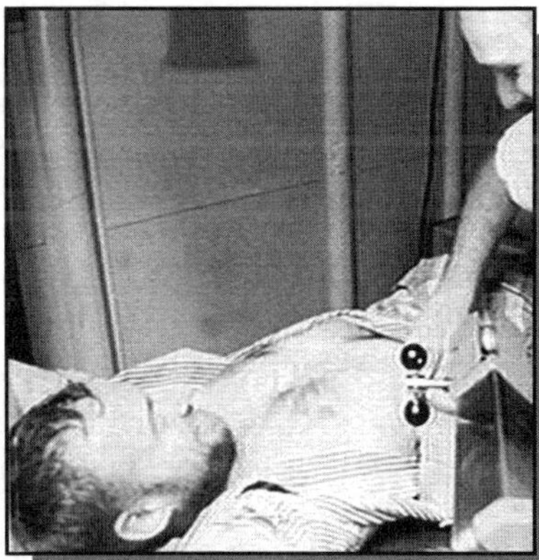

8.0 SYSTEMS OPERATION

8.1 PLATFORM

BORMAN — We aligned the platform 3 times. Each time it worked just as advertised. Daytime alignment, of course, was no problem. We got very ample yaw reference out the window.

65-HC-1169

LOVELL — Caging, for fast heat dropout took approximately 23 minutes.

BORMAN — Night time, the initial alignment is a little difficult if you do not have a full Moon. With a full Moon it is almost as easy at night as it is in the daytime. It really lit the terrain up.

LOVELL — To get your initial spacecraft attitude, the full Moon is very nice.

BORMAN — Right, without a full Moon, I think it would take you a little while to align to get your spacecraft BEF, so that you would not have to torque the platform too far for alignment.

LOVELL — You get to know the stars.

BORMAN — Yes, you have to use the stars. It would be difficult to pick up the ground and track it. Platform Modes: CAGE. Jim said that took 23 minutes for a fast heat drop out. SEF worked perfectly. BEF worked perfectly. ORB RATE seemed to be fine. We used it preparing for the rendezvous with Spacecraft 6. After running ORB RATE for approximately an hour, and then going back to align SEF, we did not notice a great amount of misalignment. The only time it was on FREE was during reentry. I guess the FREE worked fine.

LOVELL — No problem about displays, were there?

BORMAN — No, not at all.

LOVELL — Been using them for a couple of years now.

BORMAN — No problem about controls. The PLATFORM mode worked well. During our last alignment, we had degraded operation in thrusters 3 and 4; so we aligned it manually for 2 orbits. It was very easy to do, and it worked fine. We had all the confidence in the world as far as attitude reference is concerned.

8.2
OAMS

BORMAN — OAMS operational check, Pad: I think we went around the horn about 3 times before they were satisfied.

LOVELL — It took three circuits to get them.

BORMAN — Right. Inflight OAMS: The only operational check we had is when we lost the complete authority in yaw right, thrusters 3 and 4. We noticed this first in PULSE mode; we switched to DIRECT and in DIRECT we did not get, any ignition at all as far as I could tell. In the OAMS PULSE yaw right, we were getting slight little pops. It seemed we had about ¼ control authority that we had before we experienced the problem. We went to DIRECT, to see what effect DIRECT had on it and we got some thrust, but it was a whishing. We weren't getting any sound of the thrusters. It was a whishing sound. I think we were only getting an impulse either from the oxidizer or the fuel escaping.

LOVELL — We could hear a clicking of the solenoids or the operation of the valves, whatever they were back there. They were working all right, but we were not getting any resultant thrust.

BORMAN — Right. Systems Monitoring: Source pressure was fine. Went right down the predicted schedule.

LOVELL — As a matter of fact, the source pressure dropped, just as predicted, when we ran out of initial OAMS fuel before we went to the reserve tank. It came back in again when we actuated the squib.

BORMAN — No, that wasn't the source pressure, that was the regulated pressure.

LOVELL — I mean the regulated pressure, I'm sorry.

BORMAN — The source temperature worked fine throughout the flight. The regulated pressure stayed at 300, right on the money, throughout the entire flight until the auxiliary tank was actuated. We operated the auxiliary tank when the pressure dropped about 30 psi.

LOVELL — Yes. It went down to about 260 or 270.

BORMAN — Right. It came right back up, and the system worked just exactly as advertised. The propellant quantity gauge worked fine. For most of the flight it was right in agreement with the ground computations.

LOVELL — What was the final propellant quantity reading?

BORMAN — About 2 percent to 3 percent.

LOVELL — And we still had 300 psi regulated pressure.

BORMAN — Source pressure remained about 1,000 psi. Monitoring of OAMS propellant remaining: On board information I thought was good. The OAMS propellant quantity gauge, worked fine.

LOVELL — At least it was on the side favoring us.

BORMAN — Yes. The ground information was excellent. At the end of every day they gave us ground rundown of how much OAMS fuel we had remaining. It worked out fine. We were short on OAMS fuel. Any time we didn't have a specific assignment, we were in drifting flight. That's one thing we want mentioned. Every time we powered down we'd turn off the ACME bias power and the ACME inverter, and invariably this would end up in two pulses of "bump," "bump", that would tend to yaw left and roll left. And the natural tendency of the spacecraft to yaw left due to water boiler venting, I guess, and perhaps ECS venting, was aggravated by this added impulse of two blips when we shut down the ACME. How about the selector controls and switches in the cockpit?

LOVELL — No comments there.

BORMAN — I don't have any either. The attitude controller, I thought, was fine. No problems. Maneuver controllers were fine.

LOVELL — The right hand maneuver controller was a very nice operating controller and it was very handy. Very easy to operate.

BORMAN — As far as Inflight malfunctions or irregularities, we lost authority on thrusters 3 and 4. We got some of our yaw right capability back by turning off the circuit breaker for Thruster 12 and then thrusting backwards with the maneuver controller in order to give us yaw right. This worked very well and enabled us to check yaw right drift rates and enabled us to make yaw right maneuvers. The only thing – you couldn't get very small control inputs with this mode.

LOVELL — And you used a lot of gas.

BORMAN — And you used a lot of gas. I was very happy when we finally aligned the platform for reentry that we were able to get enough control out of 3 and 4 to align the platform. When we did this, of course, in order to get yaw control we went to roll jets – pitch, and that worked fine. I don't have anything to add to that malfunction. We heard the solenoids working. When we went to DIRECT we could feel we did get an impulse, but we did not seem to get ignition. It sounded more like a swishing noise. The ground analyzed it and seemed to think it was a problem with the valve seats. I'm not certain what it was. I do know that we also tried secondary drivers and that didn't help. I could tell that wasn't the problem when we first heard it.

LOVELL — We tried different modes – PULSE, DIRECT, and RATE COMMAND, but that didn't help. I think it was mechanical problem.

BORMAN — RATE COMMAND is a very tight control mode. I'm very glad it was there. I think it is very important to have that for retrofire. We also used it for reentry. I think it is a very good mode. Of course, it is expensive in fuel. We used it also for all our thrusting when we were making orbit adjust maneuvers.

LOVELL — Let me ask a question. When did you go to RATE COMMAND during the reentry?

BORMAN — I went to RATE COMMAND during reentry after guidance initiate and after I started flying the needles.

LOVELL — Because you were overshooting with DIRECT?

BORMAN — Right. I was not able to get the fine control I wanted. It would not stay in there. It seemed like the spacecraft was picking up a torque in roll also, and I was having to watch it too close.

LOVELL — And this was different than what we had in the simulation.

BORMAN — Yes. REENTRY RATE COMMAND we never used. DIRECT we used once for tracking the Reentry Minuteman in order to catch it. It was moving so swiftly. We also used it in the initial phases of reentry, and it worked fine. The PULSE mode, of course, was the one we lived with most throughout the 14 days. I thought it was an excellent mode.

LOVELL — It is a gas saver and even when you do have a platform the PULSE mode is adequate for most of the work you can do – for any attitude control, ground terrain observations – except for rapid rotations where you need a faster authority.

BORMAN — Right. All ground tracking, PULSE was adequate. We did not have any problem at all. We were able to track the Polaris using PULSE. Everything except the reentry cone we could use PULSE mode. The HORIZON SCAN mode was fine. The only thing I noticed there was at sunrise and sunset sometimes, we were driven to a 30 or 40° nose down pitch attitude by the thrusters. The scanners worked great except at sunrise and sunset.

LOVELL — They would lose lock . . .

BORMAN — Sometimes they wouldn't lose lock but, remember, they drove the spacecraft nose down. About 40° pitch down.

LOVELL — The one big thing, which was the question in all our minds, actually happened. Another spacecraft nearby will interrupt the HORIZON SCAN mode.

BORMAN — Right.

LOVELL — It does effect the scanner operation, so it is something you have to take in consideration.

BORMAN — That's right. When 6 got between us and the sun, the scanners were inoperative and lost lock. PLATFORM mode worked excellently when we had it, and we used it to align the first two times we had the platform. I think that you can do a finer job, and you can align the platform more closely manually. This is because the dead band on the PLATFORM mode is larger than you can control manually. But it certainly is a worthwhile mode and for station keeping it is a superior type of operation. Translation maneuvers at spacecraft separation at SECO + 30 – I did not hear the thrusters. I just thrusted. Jim hit the SEP spacecraft. Did you hear the thrusters?

LOVELL — No. I did not hear the thrusters. One reason why we didn't hear the thrusters in that particular case, whereas we did later on, was the fact that we had our hoods on and the air was blowing in and making a lot of noise. It was strictly by feel and by sight. No sound.

BORMAN — Right. Perigee Adjust Translation. Accelerometer bias was what they thought it was prelaunch, and it remained that way throughout the flight. This was a no platform Perigee Adjust, so, really that doesn't have any meaning there. The timing on the first Perigee Adjust Maneuver was off, thanks to me. Jim made the maneuver. We did it without a platform on a star. And, as I mentioned earlier, about this time we were in close proximity to the booster, and we started flying through some particles, but I was not sure exactly what it was, so I told him to stop thrusting as we approached this. Then, when we got in there, when we stopped thrusting, this wire came forward, hit the hatch, and I lost the timing again. We thrusted, I guess, a little too long. I am not sure exactly how long it was. I think we were aiming for a perigee of about 102, and ended up with about 120. Maybe they changed their minds and went for a perigee of 120. I don't know.

S65-63825

LOVELL — Well, that time which they gave us was not consistent with the flight plan. They gave us one minute and 16 or 17 seconds, and the flight plan called for 46, I think.

BORMAN — Well, we may get that cleared up when we talk to the ground. But, it was greater than I thought we had planned to do. Updating throughout the flight was excellent. Checklist was fine and, of course, we did not use the computer.

LOVELL — We might mention here that both Frank and I think making adjust maneuvers without a platform is very feasible. You can use the reticule for alignment and use the stars as a reference. Since you are usually using the aft thrusters, you do not have thruster light to worry about. You can turn down the lights. It takes two people though; one person to burn, hold attitude on the star, and watch the star reference and the other person to time. It required two people, but it is a very feasible method of doing it. I think you get some very good accuracies with it, because we found out from the second burn.

S65-64029

8.3 RCS

BORMAN — RCS operational checks were nominal. We had no problems at all with the RCS. System monitoring was perfect and it did not drop one bit during the 14 days. After we actuated it, it went from 3,000 to about 2,600 to 2,500 psi on the source pressure. No problems. Control modes, RATE COMMAND. As I have said, it is a very tight and fine mode. We used it during most of the reentry. REENTRY RATE COMMAND we did not use. DIRECT I used for the first part of the reentry, and it seemed that we were picking up rolling torques, and I was also starting to pick up pitch and yaw oscillations as the g's were coming on. They were slight ones but I really wanted to get the spacecraft steady, and I was really trying to lock it in on the attitude indicator, so we went to RATE COMMAND. I didn't see any reason to bring back a lot of RCS fuel anyway. REENTRY RATE COMMAND we did not use. The PULSE mode was used in the reentry prior to guidance initiate, and it worked fine. Retrofire attitude control was excellent and I'm glad we had RATE COMMAND there because we had no outside reference at all. Retrofire was done on the ball with the rate needles, and I thought the rockets were outstanding. Yeah, outstanding, I thought they were a little more powerful than I had anticipated.

LOVELL — Quite all right.

BORMAN — Reentry attitude control dead bands and rate damping was fine. The only thing, I guess, that was wrong with RATE COMMAND was the fact that it uses an awful lot of fuel. But, it certainly holds that spacecraft steady as a rock. The heater lights – we solved that problem very easily. We turned on the RCS Heaters on the second day and left them on through the entire flight. They sequenced and went on and off, I am sure, but we did not know about it. We never saw the light, and we did not have to worry about it.

LOVELL — The temperatures kept right around 80° all the time.

BORMAN — No comments on thruster firing, worked fine. We shut the RCS system down initially around 35,000 feet, shut off the motor valves and then the oscillation on the drogue built up even greater than it was. So we turned them back on again, and I'm not sure if it is my imagination or not but it seemed like this had some effect on damping the oscillations. It may have been just the position in the reentry, though. I had the feeling that perhaps we had run out of RCS fuel prior to drogue deploy. I am not certain, but if we didn't then the RCS didn't have the authority, because we were oscillating before drogue deploy. I didn't notice any RCS fumes after impact. Did you, Jim?

LOVELL — After impact? No. I think that our system of turning the Repress on and getting the . . .

8.4
ECS

BORMAN — Why don't you comment on suit mobility?

LOVELL — This was a flight that actually did some evaluation on the suit. We had the new light weight suit. Mobility is better than the 4-C suit, but mobility in the Gemini cockpit with the 5-C suit still restricts the person such that it degrades his performance for long duration missions. It is still quite immobile in the 5-C suit. We still have a lot of trouble with it. The suits checked out all right prior to the flight. We did not do any integrity checks with the light weight suits during the flight. The air flow through the suit was adequate where the flow got to the body. However, there were many pockets where the air became stagnant, especially in the crotch area. It would heat up in local areas of the body and would not provide adequate cooling. Humidity goes right along with temperature. The areas where the air flow did not go across the body, was very humid. We also noticed that it gave you sort of a wet clammy feeling when the cool air went in there. It gave you sort of a cold, clammy feeling where the flow went through. Places where the air did not reach were hot and clammy.

BORMAN — Also, the humidity in the cabin was very, very low when we were in the suits. The cabin was dry and hot. Very, very poor.

LOVELL — The humidity dew point was between 52 and 58 most of the time. We have some accurate figures on that. I don't recall any instance of even seeing the CO_2

gauge move other than during tape dumps. We had no evidence of CO_2. Comfort in any pressure suit is compromised. It restricts mobility and the Gemini cockpit is just not that big for long duration flights where you can live with the suit. Suit controls were very adequate, no problems there. We had absolutely no problems with the O_2 demand regulator. The electrical umbilical is ungainly and heavy. The connection right angle sticks out in the cockpit. It could be better designed. We did not have fingertip lights. Our mode of operation, with the suit on, was primarily with the hood off, the cover visor on, and the gloves off. Many times we also unzipped the big zipper through the crotch and up the back, and left that zipper open. We found that the big opening in the neck, with the crotch zipper closed, most of the air would go out through the neck and would not adequately vent the lower stomach area. We had planned in our flight plan to try going suitless. As per plan, about the second day, I got out of my suit and found after settling down to the environment that the skin became drier. There were no wet spots or dampness in the underwear area. I put my suit inlet hose along side of me on the center stowage area with the opening facing aft blowing air down alongside the seat blowing aft. The exhaust hose was put back into its stowage position, with the screen on, along the lower right hand footwell area. This provided adequate ventilation during most of the time. When we exercised we found out we built up quite a bit of extra heat. I would then move the inlet hose to a position along side of me, along my left leg, and tie it down along the side pedestal with the opening facing upward. This would provide more cooling into the basic cockpit area and would actually keep me a lot cooler than I had been before. We found out that without suits on, the cockpit actually became bigger. There was more opportunity to move around. You could move the body, there was less hesitancy to exercise, less resistance to exercise, you could get to things easier. You actually had more control and more comfort without the suit on. We stowed the suit on the seat, putting the visor along the outer part of the top of the seat rest and doubling the legs back against the back of the seat. We stowed the harness in the juncture of the back and the seat of the ejection seat. During zero g we were floating up and we never touched the back part of the seat.

BORMAN — I have some flight notes that I will just read out for the record. Ventilation without suits: The bypass hoses on the Gemini provide excellent return ducts for the suit compressors. They were mounted with the inlet on the outside wall near the individual crewman's outboard knee. The suit inlet hose was then positioned to secure different flow patterns. Because no provision had been made for special inlet hoses, only two positions were tried. The one most often used was the suit inlet hose located near the inboard shoulder pointing forward. This produced a flow pattern from right to left down across the body. The body was never really in the flow but a very comfortable circulation pattern was set up. The other primary pattern consisted of leaving the suit outlet hose in the same place, but putting the inlet hose near the outboard knee, pointing 90° from the direction of the outlet hose. This pattern also produced a comfortable flow pattern. In truth, I believe the cabin is so small in volume compared to the amount of air introduced by the suit circuit that almost any arrangement would provide enough air to provide efficient cooling. We also have some sketches of how this went.

LOVELL — We also believe, after spending several days without suits on that the theory that at zero g there would be no convection cooling . . .

BORMAN — I am sure that there isn't any, due to change in the heating condition.

LOVELL — The mass of the air being pushed out by the compressors is enough to give adequate flow throughout the entire Gemini cockpit. We had no problems with air flow.

BORMAN — It would have been a very, very difficult task to stay in those suits for 14 days, if not impossible. We certainly would have been in much worse shape when we got down.

LOVELL — I believe so. We were requested from Houston to try the hose position evaluation whereby the inlet and outlet hoses were together.

BORMAN — Yea.

LOVELL — We tried it and to be perfectly honest, with the small cockpit and the amount of flow out of the inlet hose, we did not find much difference, it was adequate, but it was awkward to use it that way. We did find out that if my exhaust hose was put on Frank's side that we would get stagnant spots on my side of the cockpit where although I wasn't uncomfortable . . .

BORMAN — . . . that is with your inlet hose being turned off. So all these flows were introduced on my side and both the return hoses were on my side.

BORMAN — That is right, they were on your side. I found stagnant areas, I wasn't uncomfortable, but I did find stagnant areas where there was no flow going on on my side. You have to have adequate positioning of the exhaust and inlet hoses.

BORMAN — I think really to solve the whole problem, if you want to design efficient cooling for suits off operation in spacecraft, design it the same as you would on the ground.

LOVELL — Right, I think you are right.

BORMAN — You would have no problem. For instance, in most of the airplanes now that are pressure cooled the flow is so great across them that you have a continuous flow pattern in there, coming out usually from one inlet located up around your right shoulder and a couple in around your feet. I don't think you have to worry about the bugaboo of no convection in zero g because it is overshadowed by the large kinetic energy input through the large amount of air.

LOVELL — Cabin pressure was 5.5 on lift-off, came down to 5.1 and stayed there exactly 5.1 for 14 days. It did not move when we jettisoned the adapter and went on the bottles, it stayed exactly 5.1. The only time I saw it move was on the water when we used all the oxygen up and it went to zero.

BORMAN — Well, it was below 5.1 when we opened the Inlet Snorkel.

LOVELL — Inlet Snorkel . . . yes, that is right.

BORMAN — You were talking about the inlet bottle pressure.

LOVELL — Yes, that's right, the bottle pressure.

BORMAN — The temperature varied with the suits on and the suits off operation. I have gone through my notes here, and I note that it says when we both had suits on and we were just barely cool enough with both suits on and the B pumps running. On the other hand, when we were both out of the suit, the B pumps only running, we were very comfortable. When we were up working and operating, we noticed that the temperature level in the cockpit was just right. We were running most of the time with the suits full cold, the heat exchanger full cold, and maximum air flow on both controls. Then for several days when we went to bed, we left the controls that way, and we would wake up very cool.

LOVELL — Well, there are several factors. I think the size of the Gemini cockpit and the fact of a completely closed loop system is very dependent on 2 factors. One was the heat output of the people and two was the amount of heat you get in through the windows due to the sun. At night our heat output decreased, we put up shields on the windows, including some aluminized foil to reflect the sun, and I think the combination of both these things with the systems we had during the day time really dropped the heat in the cockpit. Then, during the day when we were active and had the windows open again, the temperature increased inside the cockpit so it was very comfortable.

BORMAN — That is right but the last couple of days we turned down the suit flow at night and it helped out.

LOVELL — To compensate for this thing we turned down the suit flow.

BORMAN — There's a lot of inertia in the cooling system and it takes a long time from the time you make a move on the controls before you can feel it.

LOVELL — Just to regress here one minute. When I was out of the suit and Frank was in it, we put my suit flow to full decrease and his to full increase to give him maximum cooling in the suit. I was not uncomfortable with the full decrease flow in the cockpit.

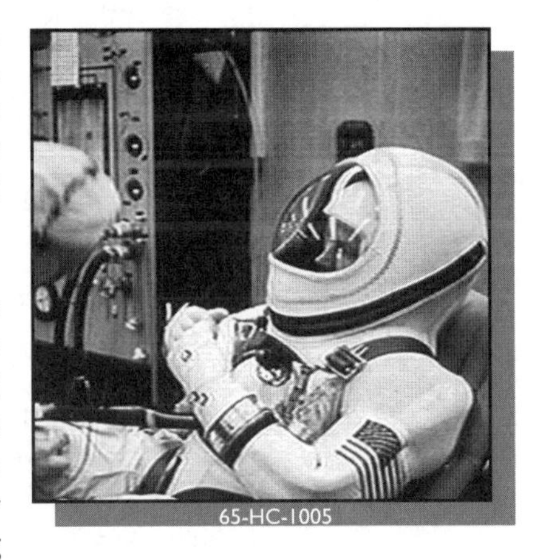
65-HC-1005

BORMAN — The humidity in the cabin was higher with the suits off. It was a much more comfortable cabin. Your skin didn't get dry, and the nose problems we had the first 3 or 4 days went away. I am not sure that we can contribute this solely to being out of the suits or whether it was the fact we were becoming more acclimated to the 100 per cent oxygen.

LOVELL — We have some accurate figures. I believe though, that with suits on the humidity-temperature range was about 20 degrees difference. With the suits off they went around 10 degrees, I suspect.

BORMAN — We have them all there. The only time the CO_2 jumped at all was during tape dumps, it would go up and then come back down but we knew this before flight though. Comfort day and night, with and without suits . . . There is just no comparison. I have the notes that I wrote down while we were still up there. There is no comparison between suit on and suit off operation. The suit off is 1,000 per cent better. I think I may have been conservative. It was maybe a lot better than that. Comfort without the suits was by and large very good. We used the cabin fan only once in the entire mission during one of the checks with the suit off. This was when we had Jim's suit inlet hose blocked off and my inlet hose operating in my side and the two suit outlet hoses in my side of the spacecraft. As we already mentioned here, Jim noticed some stagnant areas in the spacecraft, and we turned on the cabin fan to see if this alleviated the problem, and it did help. There was a definite circulation with the cabin fan on. The only problem is the cabin fan draws a considerable electrical load, and we did not have the power to run it continually. The cabin pressure regulator worked perfectly, it never worked at all.

LOVELL — That is right, never heard it, thank goodness.

BORMAN — That's the pressure relief valve. Right, never heard it. It never actuated. The cabin pressure regulator was as steady as a rock. As Jim has already mentioned it stayed 5.1 the whole flight, and I never saw it budge at all, until you opened the snorkel.

LOVELL — Right.

FCSD REP — Cabin vent valve.

LOVELL — We had a double vent valve with the tip bent up to protect the stop. We never used that until we got down to the checklist during the reentry portion of the flight.

BORMAN — The Cabin Repress Valve was on the entire flight because we had the M-1 Experiment hooked to it. Then, of course, we actuated it again when we got on the water just to get some cooling oxygen into the spacecraft. I have no comments on it, we had no problem with it, the friction had been increased on it so that it worked quite well. It stayed in the open position for the entire flight. The Cabin Air Inlet Valve, we used . . .

LOVELL — Just during the reentry phase.

BORMAN — Right. Just during the reentry phase.

LOVELL — . . . with the snorkel valve, that is when I think I got a whiff of that stuff through there.

BORMAN — All the time with the suits off, we were running with the Cabin Air Recirculation Valve closed. The rest of the time when there was one person in the suit and one person out of the suit we ran with it in a 45 degree position. When both people were in the suits, we ran it in the 45 degree position.

LOVELL — I'd like to make one comment on the Cabin Inlet Valve, I think a future procedure would be to either open the visor or unzip the hood prior to using the snorkel valve, so you do not get this concentrated ambient flow into the suit in a small concentrated area. Okay. Primary O_2 System Monitoring.

BORMAN — Primary O_2 System Monitoring was no problem. The cryogenics bottle for ECS O_2 oxygen did vent . . . I think it was about the 8th day it started venting. Performance of the cryogenics bottles have been outstanding. The first one, ECS O_2, started venting today. And we were still adding heat to the other two bottles. Lets see, this was Sunday morning so that would have been the 8th day that it started venting. The quantity measuring system worked fine. The flow rates were adequate. I just cannot emphasize enough the desirability of going without suits. The pressure and temperature remained just nominal. We had dome plots aboard the spacecraft, and we checked them out. We had to use the auto heater on the ECS O_2 bottle perhaps the first two days or so and then we were able to turn the heater off completely. That bottle had a big enough heat leak so that it maintained pressure itself. As I said, it started venting on the 8th day. We never used the manual heater on the ECS O_2 bottle.

LOVELL — As a matter of fact, the primary O_2 helped, rather we utilized the primary O_2 to pump up the pressure on the FC O_2 sometime in the early part of the portion of the flight.

BORMAN — Yes, we used the crossover valve. When you hit the squib or when you hit the switch to open the solenoids, even though we had been led to believe that it takes some time for that pressure to build up, it looked to me that it went to about 250 in the FC O_2 bottle almost immediately. It went from 100 to 250 almost immediately. I imagine it will come down when we talk about the FC O_2 problem.

BORMAN — Secondary O_2 System Monitoring was nothing, we checked it . . . all the time, but GO / NO GO decision once a day. It stayed exactly the same throughout the entire flight – 5400 and 5300.

LOVELL — It did not budge at all.

BORMAN — Quantity measuring. We do not measure except for pressure, and as we said that stayed constant. Flow rates, pressures, and controls were nominal in the secondary O_2. You could not tell when we had gone off the primary onto the secondary O_2. We have already talked about the CO_2 partial pressure. It was below zero the entire flight except during tape dumps when it jumped up due to a glitch that the tape dump puts in the TM system. Radiator operation and configuration. We ran radiator on all the time except for two checks that were made. Actually when we open circuited the fuel cells before we brought them back on the line, we went to RADIATOR BYPASS twice. Then we went to RADIATOR BYPASS once when we wanted to get the water out of the ECS System. One time during the flight we were picking up water. This might be a good place to cover that. We were picking up water coming out of our suit inlet hose, in quite large quantities. We called the ground, and they suggested that perhaps the water boiler was not venting. They called up the procedure that included putting on evaporator heat, turning off the radiator and going to bypass on the radiators and rotating the spacecraft at 10 degrees per second. This threw out a large amount of water and things got back to normal. Later on in the flight we noticed the same thing, but we were busy aligning the platform and other things and we did not want to setup roll-rates. All we did was put the suit coolant to warm and put both suit fans on and blew the water out of the system, and that worked also. We did notice that during the flight, down in the vicinity of my right foot in the center pedestal lower area there, around the cabin heat exchanger, we got a lot of condensation. It was very wet on the walls and the blotter paper was sopping wet, a radius of about 12 inches on the lower right pedestal, on my side. I do not know how it was on your side.

LOVELL — It started getting wet on my side, the inboard.

BORMAN — Yes, inboard side. That is the only place in the spacecraft that I noticed any condensation. As far as I know the evaporator only operated during launch and the first orbit. We also had one other day when we woke up and were tumbling quite badly and the wall temperatures were 16 degrees to 20 degrees lower than we ever recorded them before. I suspect that the evaporator might have worked that night. We mentioned this to the ground and . . . let's see, I have some notes on that.

LOVELL — It probably got filled up from the moisture going into the system.

BORMAN — This was noted in the data for the cabin temperature surveys also. I don't remember exactly what day it was, where we noticed this big change in temperature.

LOVELL — It was about 5 or 6 days after the mission started, wasn't it?

BORMAN — Yes . . . here it is here . . . it was 158 hours and 27 minutes when we got up, and we had a wall temperature of 64 degrees and a pilot hatch temperature of 66 degrees. Comparing this with 144 hours and 53 minutes, the hatch temperature had been 84 degrees, so there was a 20 degrees drop during this one evening. I attributed this, plus the fact that we noticed large drift rates when we woke up, to the fact that the water boiler must have been operating during the night. It was the only time during the flight that we noticed these large drops in temperature . . . the very cold wall temperatures. We were on double loop, B pumps most of the time. Finally, went to A pump twice in the flight when people were in the suits in order to stay cool. We went to double A pumps, of course, when we were powered up.

LOVELL — We had one time when we had one A pump on.

BORMAN — Twice we had it on.

LOVELL — Twice, but just one primary pump. We did not go to two A pumps.

BORMAN — No. A pump in the primary loop was on twice to keep cool.

LOVELL — Right. The secondary loop A pump was not on.

BORMAN — Never on except during periods when the platform was running.

LOVELL — Right.

BORMAN — Now, as I mentioned before, when we were running both B pumps with the suits off, it was comfortable. When we got the suits on, and both B pumps going, it was not enough to handle the load.

LOVELL — It was marginal.

BORMAN — That is right. Normal mode was all we used on water management.

LOVELL — Never touched the . . .

BORMAN — Never touch the condensator.

LOVELL — The drink gun worked as advertised.

BORMAN — One thing I would mention is the fact that this logging every ounce you drink was an operational nightmare.

LOVELL — I think the gun is adequate for flights if all you want to do is to know the total quantity of water that is going out for ballast purposes or CG purposes. I do not think there is a requirement to know just how much each crewman is drinking as long as it is adequate. There is no need to log, all you have to do is report counter readings once a day for the guidance people and fuel cell people to know just how much water is being consumed.

BORMAN — Right. It says flush mode. We never used the Flush mode or Evaporator Fill mode.

8.5
COMMUNICATIONS

LOVELL — The interphone Operation and Quality was okay, without the hood on. The G5C suit made communications poor, because of the flow of the air to the hood. Other than that I thought the interphone was pretty good.

BORMAN — Yes, we should mention the fact that the G5C suit with the hood zipped did introduce a lot of noise.

LOVELL — Yes, a feed back into those two mikes and there was a lot of noise.

BORMAN — The quality of the interphone was excellent. With the suits off we didn't use them most of the time, just like talking in a room, so we didn't need it. My UHF was a little fuzzy during countdown.

65-HC-1002

LOVELL — Mine was good.

BORMAN — Yes, yours was good and mine was a little fuzzy. I could hear people all right but they claimed that I was a little weak. In orbit, I just can't say enough nice things about the UHF.

LOVELL — UHF was excellent in orbit for the entire 14 days. Very little static. High quality reproduction.

BORMAN — With the squelch on zero.

LOVELL — Right, the squelch on zero.

BORMAN — The UHF performance during recovery was excellent. No problems with that at all.

LOVELL — We did have trouble getting back to Houston.

65-HC-1042

BORMAN — I was very pleased with the entire voice procedure operation around the world. I thought they did an excellent job. We didn't have any problems at all. They were quiet when they were requested to be during our sleep period.

LOVELL — They were outstanding.

BORMAN — They were outstanding, yes.

LOVELL — The voice tape recorder operation was fine. There were no hitches as far as operating the voice recorder. It was easy to, well, that's three feet of change. We used mostly the CONTINUOUS mode rather than the MOMENTARY, I used MOMENTARY when we just wanted to make a comment. As a matter of fact, I think the MOMENTARY position does save a lot of voice tape, because you don't have it on and forget it. However, we had a procedure with the voice tape that was going to record the quantity of urine that was dumped. And this led us to leave the voice tape on quite a long time when we weren't doing anything or saying anything – and using quite a bit of tape. I think that it would be helpful if we had some sort of a little light of some sort to let you know that the tape is on. When we have a flow meter which was being evaluated for future flights and might be a standard piece of equipment then it would certainly be nice to have some indication if the tape is on. On long flights you can't have the tape on all the time like the short flights – you have to conserve tape. Cartridge change was no problem. The controls were adequate. Data Recording? We tried to record as much data as we could.

BORMAN — We didn't indulge in a luxury many times of recording both in the log book and on the tape. We only had 20 tapes for a 14 day mission. If we got a good representation of it in the log book – we didn't put it on the tape. Now, we find out we probably brought back some unused tapes, too.

LOVELL — We did. It was hard for a 14 day mission to adequately budget the tape. We would try to budget it so that we could get the information on there without leaving long periods of inactivity on the tape. However, we didn't budget it well enough, and we left about 1 or 2 tapes without any recordings.

BORMAN — Digital Command System updates were good, no problem – everything worked fine. Real-time transmitter, and delay-time transmitter were no problem. As a matter of fact, that whole system I thought was excellent. The only problem we had in the area of Digital Command System or the telemetry was that we lost the tape recorder and . . . goofed up the delayed time. The procedure that we worked up for operating with Spacecraft 6 in the air, I thought went very well. It posed no problems. Communications Controls and Switches – Voice Control Center, Audio modes, Keying and Antenna Selection, were all nominal. We might mention in Sleep Configuration – we never used the Sleep switches because we had the situation where we pretended it was night and went to sleep every evening and the ground never called us. I don't think they ever violated that for the 14 days. They never called us during the sleep period.

LOVELL — So that worked very well.

BORMAN — Beacon Control was no problem. We didn't use the Reentry C-band Beacon until reentry. The TM controls, transmitter, and antenna again were no problem. It was operated just as advertised due to all instructions from the ground.

8.6 ELECTRICAL

BORMAN — Now we have some interesting things to talk about.

LOVELL — Well, we monitored the electrical system pretty closely.

BORMAN — Yeah, I guess we did.

LOVELL — The only thing we can say here is reiterate what we have probably said before. On lift-off we had delta P lights come on for fuel cells – both sections. I blinked on and off several times and went off. 2 blinked on and off several times and stayed on through insertion and stayed on most of the time during the 14 days. We have recorded in the flight book of the flight plan – those times that it went off and on to the best of our knowledge. I'm sure we missed several of them.

BORMAN — When we were sleeping particularly.

LOVELL — When we were sleeping we missed them, but it appears to me that there are two things now that these fuel cells have a lot more latitude than we really first realized. We can operate with the fuel cells with delta P lights on more than we thought we could. As a matter of fact, we were doing normal purges with the fuel cell delta P lights on which the systems book said flatly not to do. But we had some excellent guidance and assistance from the ground in keeping the sections running.

BORMAN — I think so, too.

LOVELL — I think that's what kept Stacks 2C and 2A going as long as they did go. The gauge is a little inaccurate to monitor the system. If we are going to have troubles with fuel cells as we did on this flight, and if the ground is going to keep requesting accurate stack amp readouts. The gauging system is poor because it is hard to read accurately the amps when they are down in the low 1 and 2 amps. Each indicator is canted a different way – alternately throughout the 6 stack readouts. The ones that are canted inward away from you are hard to read.

BORMAN — The fuel cell, as Jim said, was an interesting thing. We finally lost stacks 2A and 2C about the 11th or 12th day. Stack 2B remained on and I'm sure there is a whole history written on the ground of the things they did and tests they ran at McDonnell when we were in the air to see just what they could do and how far they could go with these fuel cells. I thought they did an excellent job, and we ended up being able to run them the whole time. As a matter of fact, we turned on our Squib batteries about the 10th day – used the Bus Tie switches and were running entirely on the fuel cells the latter part of the flight.

BORMAN — The onboard cues for monitoring the electrical system are adequate. We found out one thing in this flight, that is the Delta P lights really don't mean a lot. We had been told before the flight never to purge if you had a Delta P light. We ended up violating every single one of the cardinal rules that we had.

I think the thing to note about the entire electrical system was the fine work done on the ground. They came up with solutions. They evidently were running similar type cells at MAC, St. Louis, and they kept them working for longer than they should have.

Fuel cell operation, as far as I was concerned, Section 1 was ideal. Section 1 maintained its share of the load the whole flight. Section 2, we lost stacks 2-A and 2-C eventually, I believe on about the 12th day. I was a little concerned on the 13th day with the status of Section 1 because we had had a delta P light on Section 1 for the first time and we had been running almost 24 hours. But, the ground came through and read us up a technical report from St. Louis that explained the whole thing. It made me feel a lot easier when they did that. Rather than having the ground comment blindly on it, "the fuel cells are going to be good for 24 hours," I would like to get a little background information on it. How else could we know it was going to be good for 24 hours, and what had they done to prove it would be good for 24 hours? They read it to us over CSQ, it eased my mind a lot because I wasn't anxious to miss the WASP. On the 13th day, I wanted to be able to go the full 24 hours rather than have to land in the Pacific. So the whole story of the electrical monitoring, as far as I am concerned, was great work by the ground.

The main batteries held constant between 22.5 and 22.7 amps for the entire mission, and we checked them once a day at the GO / NO GO stations. When we turned them on 2 hours before retrofire they carried their share of the load and were operating fine when we were in the water. We turned off the squib batteries about the 10th day and used the bus tie switches. We ran entirely on fuel cell power for the last five days. When the squibs came back on, the voltage was 25.5 after they had been turned off for five days. They operated properly for the last 2 hours of the flight.

8.7
ONBOARD COMPUTER

BORMAN — During the launch it was absolutely a nominal case. The pitch status, yaw status, and roll status, were zeroed except for a brief period at guidance initiate when they went out about 2 to 3 degrees, and then zeroed. We had no violent pitch down at guidance initiate. Attitude indications were nominal all the way through. At insertion, the nominal velocity on address 72 was 25,804 and when we read it up, it read 25,804. The orbit maneuvers using the computer and the platform were right on the money. The accelerometer bias did not vary, and we burned them off on the IVI's by inserting them through the MDIU, and it came out very well. I did not burn on time, we burned on the IVI's.

BORMAN — The updates were all made in the PRELAUNCH mode as agreed on before flight. There was no problem, no misunderstanding, I think FOD did very well in this regard. I know that in Gemini 5 there was a little mix-up, but we had none of that.

BORMAN — Retrofire occurred automatically at the exact second. All four Retros fired and the IRS was right on the money. Reentry guidance was nominal. It was very similar to the simulations that we had flown. There was one little anomaly in the guidance, in that we were given back-up reentry angles of 50 degrees. We computed with our onboard charts a reentry angle of 50 degrees, back-up angle of 50 degrees. But then after tracking, the ground called up 35 degrees which proved to be closer to what we actually flew. I am still not aware of the reason for change; why it changed from 50 degrees to 35 degrees. The important thing is that it did change and the ground was able to update us in real time, and it agreed very well with the actual case.

BORMAN — The MDU worked perfectly the entire flight. Computer modes, PRELAUNCH, ASCENT, CATCH-UP, RENDEZVOUS, REENTRY, were all perfect, no anomalies in any of those.

8.8
CREW STATION

BORMAN — Controls and displays. The sequential telelights operated exactly as programmed. At minus 2:56, they came on to the second. They all turned green when they were punched, no problems there. The event timer was used only intermittently throughout the flight for timing, and for the last 20 minutes. It worked fine. The IVI's also worked exactly as planned. The Flight Director Indicator was again, a nominal case. One slight difference between the simulator and the Flight Director Indicator in the spacecraft, was the little outer roll gimbal indicator in the simulator always came up to the top. I'd grown used to flying the reentries by using that as a lift vector. In the spacecraft when we got all set up for reentry, lo and behold, the outer roll gimbal was down at the bottom, so I had to fly the reciprocal of it. But it was just a minor change and I ended up acclimating to it with no problem. I think it is just a function of how you happen to go through zero. If you go through zero just a little bit to one side, the gimbal goes to the top, and if you go through the other side, it goes to the bottom.

BORMAN — GLV fuel and oxidizer pressure gauges were nominal. The concept of sticking the decals on the outside of the gauge is poor at the best. But, we all know this has been done, and they're not going to change the gauges, and it worked fine. I would suggest never going this way again. I think we ought to change the meters in the future.

BORMAN — The altimeter worked fine. The only problems we had with the altimeter was when we were oscillating violently on the drogue, it was not indicating descent. As soon as the drogue oscillations steadied out, the altimeter came down very well. Rate of descent indicator was likewise. As a matter of fact, I can't tell you what the rate of descent was after we opened the main chute. The main chute was so good when we looked at it. We didn't see any gores or frays. And when we went to single point release, I didn't even look at the rate of descent indicator. Did you?

LOVELL — I couldn't see it.

BORMAN — Did you even think about it?

LOVELL — No, you mean to tell me you didn't look at the rate of descent indicator?

BORMAN — We could tell from the altimeter we were going down very slow. The accelerometer seemed to give us slightly lower values than the recorded. I think on the reentry the highest value we got was 3.9 g's. During launch the highest that was recorded on the accelerometer was about 6.75 g's. I understand that the actual value was over 7. On the nominal profile, it is.

LOVELL — Was the reentry a little higher than 3.9?

BORMAN — I don't know, I doubt if it was, it was so near full lift. Switches and circuit breaker panels. We had a couple of cases knocking off circuit breakers. We did have one fuel cell control circuit breaker pop on us twice.

LOVELL — I am not real sure it popped. I don't know whether I hit it inadvertently twice.

BORMAN — No, you didn't. The second time I watched it pop. Other than that, I thought the switches and circuit panels were well located. I think it is very important that we have those guards on there, particularly with changing suits.

65-HC-1064

LOVELL — The fuel cell switches, the power and control switches, should be LIFT to move switches. They should be over center locks that you have to lift to move them up. There was a guard over it, but still it was so easy to reach up there and hit those things. I was always worried about throwing the control switch off, which would have really fouled up the fuel cells.

BORMAN — You mean like the squib switches?

LOVELL — Yes, like the squib switches. I think that is the way they ought to be because you never touch them unless deliberately.

BORMAN — Yes, that is a switch that is never moved unless there is a failure in the fuel cells.

LOVELL — They should be a little better type of switch than they are.

BORMAN — Mirrors. Operating without suits on, I found that I seldom needed the mirror. I don't believe I used it more than 2 or 3 times except to check and see how far my beard had grown. How about you, did you?

LOVELL — Well, they were good for things like looking way back in the corner, and shining a light back there.

BORMAN — I was mobile enough that without a suit on I could turn around very easily and see all around. There is no question you need the mirrors. I am not suggesting even remotely that you take them out. With the suit off it cuts down the need for them. The swizzle stick we used once to pry up the center line stowage bracket. When we opened the center line stowage after launch, the bottom bracket sprung down about ¾ inches, and we had great difficulty to close it. We only closed it twice during the flight after that. We just kept it velcroed partially shut.

LOVELL — I think the boost acceleration sprung it out of position.

BORMAN — Either that, or when they put that fix on there to beef it up, it resulted in an out-of-tolerance situation. I hope that the people did not force it shut and then let us take off that way. That was a pain in the neck to get it shut.

LOVELL — We had to use the levers of the swizzle stick to get the thing back together again. This was bad. We also used the swizzle stick to keep the manual heater switch down on the FC H_2 which is a real big pain. It is a very small switch and you have to hold it for a long time. That gets to be a lot of trouble.

BORMAN — Lighting, indicators and instruments. There is one instrument in the spacecraft that should be lit that is not. That is the digital timer. That is the most valuable instrument onboard. We used it continually, it never varied one second in 339 hours. We never had to reset that once. It was exactly on the money. We checked it periodically and it never gained or lost a second.

LOVELL — But it had to be lit.

BORMAN — It should be lit because it is right on the center panel, and there is no lighting on it except for the bright light from the back. Many times at night and when you are trying to maintain dark adaptation you end up having to use the flashlight on it.

LOVELL — It should be a red light for night work. In the day time you don't need it because the cockpit is lit up anyway. It really ruined night vision to turn on that flashlight to find out what time it was.

BORMAN — The left panel was fine when it was lit up with the display light. The center panel and the right panel were all right. There is no question that the lighting system on the LEM is superior.

LOVELL — We used the red lighting more than I thought we would ever use it. We never used it in simulations. The red lighting turned out very nicely when we started looking out the window, using the stars, getting oriented and things of this nature.

BORMAN — The pedestal, console and circuit breaker areas just aren't lit. Same way with the water management panel, when you wanted to check that, the only thing you could do was use your flashlight. It was not a big problem. The little flashlight that CSD developed, and put in, was one of the most valuable pieces of gear we had. We used it continually throughout the flight, it is much more valuable than finger tip lights. I see no reason for fingertip lights because you're not going to fly with gloves on most of the time. If we would have had them, they would have been stowed. This little flashlight turned out to be a little jewel. The utility light I did not use once in the whole flight.

LOVELL — I turned mine on once to see if it worked.

BORMAN — The flashlight was much easier to get to. We velcroed it right in front of us, and it was very handy. And at the end of 14 days it seemed as bright as it was before. One of the serious deficiencies in the flight was the dirty window.

LOVELL — I just talked to John Brinkman about the film. He said a lot of it was good, but a lot of it they could tell the window was dirty.

BORMAN — What about the booster film?

LOVELL — They haven't processed two rolls of film yet and they don't have the Polaris launch yet. The Houston one turned out sort of hazy and I thought it was a clear day. He told me there was haze on the ground.

BORMAN — Did they see Houston all right?

LOVELL — They could make out the International Airport.

BORMAN — You took that with the 250 mm lens.

LOVELL — The high speed film. They don't want to process it until they talk to us to find out what kind of exposures we used. We had all kinds of exposures. I think the picture that was in the paper was from the 16 mm camera. The window was very dirty. And I have a . . . a picture was taken of it. Shows the . . .

BORMAN — Jim drew a sketch of the window in the S-8 / D-13 log.

LOVELL — There was a greenish, greasy film over the whole thing right in the center. Outside of that was a sort of a haze or fog effect. Right along the outer edge, it was clear. If I focused on the nose of the spacecraft it would be blurry. Just off the nose it would blur out. There are two theories, one group of people say it's the nose cover that is ablating on launch, others say it is staging.

We saw quite a bit of flame at staging and it looked like there were several streaks there caused by staging. There is also a general deposit like a stagnation point right there that might have been built up during the entire launch, which might be the nose cover. So, it might be the combination of both.

BORMAN — It might have accumulated due to the urine dumps throughout flight. Several times we saw urine crystals come back and hit the nose cone. We never saw them actually hit the window. I am not sure that some of it, that was practically invisible, might have hit the window. It did seem to get worse with flight. My window was not nearly as bad as Jim's.

LOVELL — Frank's was better than mine. Whenever I could I would give him the cameras to take a picture. He did a lot of the Apollo landmark and S-5 and S-6 pictures while I was controlling the spacecraft.

We have to improve the windows somehow. We've got to have some sort of cover or get some certain type of material. The windows were perfect when we got in the cockpit. The problem they had on GT-5, where they had fog and humidity because of the difference in temperature when the White Room was dismantled, was not there this time. Intensity control was good, no problem. We had two white lights in the center cockpit, this was our request a long time ago, and after using it I think we made a mistake. We never did use the thunderstorm light that we stuck in place of the red light. Right now Frank and I think we could have used the red light again because we both did use red lights a lot more than we thought we were going to, for night work. It gets your eyes accustomed to the night, and you can see the airglow and stars a lot better. If you have bright lights on in the cockpit, at night, with glare off the window and your eyes adjusted to the white lights, you could never see out. It's just black.

Onboard data: checklist cards preparation, excellent. I think the people who made them up, Chuck Stough, has to take a personal bow because I think that he did an outstanding job of making up all the onboard data books and cards. They are very, very good. What we did was, tear off the lift-off cards prior to reentry and just had the reentry section, so we wouldn't get mixed up. There are several minor things which we could change to make it a little bit more compatible, like getting one card with all the data on it so we didn't have to flip the cards back and forth when MCC gave it to us. As a matter of fact that is exactly what I ended up doing, I took the core card, and after I read the various cores for the reentry parameters, I got the nominal IVI's, also the bank angle updates and things of this nature, all on one card. Then I went back to the other section and transposed them in there.

BORMAN — One of the most important things about the checklist on this flight was the fact that we had them about a month before the flight. We used them in training, and the people responsible for that did a great job, Chuck Stough and Ted Guillory.

LOVELL — That is important. On GT-4, because of the newness of the system, we were still rearranging cards and books just prior to the flight. Learning from that flight, on GT-7 we really gained a lot by having the cards and books early in the game so we could train with them. Checklist cards usefulness was outstanding.

BORMAN — The maps and overlays were fine. We carried the larger orbital display map. I'm not really sure we needed it. It was a little cumbersome in the cockpit. It was all right, but general areas would have been just as readily available on a small map. When we were doing the Apollo landmarks, particularly those with coastal features, I thought Apollo landmark maps were entirely adequate. I did not see any reason for photographs. If you really want photographs, the best way to do it would be to fly over them with an airplane and then change the scale to whatever you wanted it.

LOVELL — The photographs were important, but I don't think you have to spend valuable fuel and time to get them. An airplane can do the same job getting photographs that we need for Apollo landmarks.

BORMAN — That's right. No question about that.

LOVELL — I found it difficult to move the map overlay.

BORMAN — It got better as the flight went on.

LOVELL — Yes, because we wore it in. The overlay we have, with periods, orbits, and the map underneath, I think that can be improved. We needed a very simple device with two rollers on the end, or some system a little bit more elaborate, but a lot easier to handle.

BORMAN — I don't know, it worked all right toward the end, Jim. If you get it too elaborate, or too easy to roll, then it is going to change on you.

LOVELL — It has to have a system where it can't change.

BORMAN — It was valuable. You knew where you were all the time.

LOVELL — We used it more than we used the star charts. Mainly, we used the star charts for the no platform burns, for retrofire position, and for SEF and BEF positions.

BORMAN — By and large the maps and overlays were well prepared. They were available early to us. We knew how to use them and it was a very, very fine job by FCSD people responsible for them.

BORMAN — Data books: We were using a system that was started in GT-4, furthered in GT-5 and I think it is working out very well. If there is any derogatory remarks on it at all, it is the required amount of logging you have to do. It is really a double entry system. But, hopefully this will cut down the postflight activities and give people a better idea of what they are looking for. I would not suggest even for a moment that we change it. We did delete some of the redundancy to endeavor to save voice tape. We tried to log everything in the book, but many of the things we did not put on voice tape that were already in the book because we wanted to save the tape. We only had 20 tapes for 14 days. Everything that was done is in the books. Most of the critical things that were time significant are on the tape.

S65-63826

BORMAN — Star charts, Polar and Mercator. We used the Mercator almost exclusively.

LOVELL — I'm not saying the Polar was not any good, but the Mercator was very adequate and we knew how to use it. I enjoy that particular type of chart a little bit better.

BORMAN — It was preference more than anything else. We did check the Polar out during flight and it was apparently accurate. I didn't see any reason to change those star charts either, did you?

LOVELL — No, I think we have enough stars on there. I think they're adequate.

BORMAN — Stowage at launch was a little gruesome. When we got in we found all the stuff stuffed on the floor over our feet. Once we got into orbit and started going through our prearranged procedure there was absolutely no problem. We used the food bags to put the refuse from each meal in. We usually stored three meals in the front until we were ready to dispose of them, and then we would put them behind the seat. We filled the debris guard areas we had behind the seats in about eight days. After that we stowed the used ones in the bags we had. For reentry, we placed them over the seats as we had done before. It worked fine.

S65-63784

LOVELL — When we first started training for the flight, there always seemed to be a de-emphasis on exactly how much we were going to stow. For instance, the size of the food bags was a lot smaller than it turned out. The size of the tissue we used was a lot smaller. I think that we ought to look at it realistically early and make sure that we get the right sizes. We were led down the path there on that first stowage review in St. Louis.

BORMAN — Yes. We caught up with it on the third one though. We doubled the size of everything.

LOVELL — That is right.

BORMAN — We took an actual meal and ate it and got the refuse.

LOVELL — It was very fortunate that we did this. It caused us to look for new places to stow things.

BORMAN — As it was it worked out real fine. The cockpit was cleaner when we reentered than when we left. Another item that was very helpful from the cockpit cleanliness standpoint were these by-pass hoses with the screens on them. They acted as vacuum cleaners on the whole flight. All the garbage and refuse would get collected on them. We could clean them off and put them in the bag and it worked great.

LOVELL — We never had any large amounts of dandruff or anything floating around.

BORMAN — The harness we took off. All you can say about the harness is it is a necessary evil. Once you get it off, it is tough to stow. Jim, you sat on yours, didn't you?

LOVELL — I stowed mine between the seat and the back of the ejection seat because it was a dead space for me.

BORMAN — I stowed mine on the outboard side of the seat. We never took the life vests off the harness.

BORMAN — Waste disposal and stowage. We used the aft food boxes for the defecation gloves and the urine sample bags. Jim filled up his first and then I started filling up mine. We ended up putting one day of food and some other refuse in the left-hand food box in addition to the defecation gloves and the urine sample bags. One thing we might note is the horrible odor every time we opened those boxes to put something away.

LOVELL — It was a necessary evil, Frank.

BORMAN — We were a little concerned when we opened the vents on the boxes for the reentry that the smell might be with us for a couple of hours, because we had to open them early before we put our suits on. But evidently the vent is just fine. It is large enough that it equalized the pressure, but it is not large enough that it lets the odors escape into the cockpit.

8.9 BIOMEDICAL

BORMAN — Oral temperature measurements and thermometer, no problem. Although, it seems strange to me that we have to have a TM temperature. That thing got in the way.

LOVELL — Yes, the tube got in the way and floated around, and you almost poked your eye a couple of times with the thing. It is a thin probe. It is very awkward because there are two of them. One is in the lightweight headset, and if you do not have the suit on, you have to stick it down through here. If you have the helmet on, it is supposed to be sticking out here, and it gets in the way. If they want Inflight temperature, we should take along a regular thermometer. We had a lot of glass in the cockpit. I do not see why we cannot carry some sort of a plastic thermometer. It seems ridiculous to me to have to TM a temperature.

BORMAN — I must admit I did not even know I had a blood pressure cuff on except when I filled it up. It did not cause any skin problems or anything. It is probably as good a way as we can go.

LOVELL — It seemed like I pumped up my cuff a lot more inflight than I had to on the ground for the same measurement. Sometimes we would not get the comment that, your "cuff is full" from the ground until after your arm was quite puffed up. Some times your arm really got to be sore. I do not know what you can do about it.

BORMAN — We used the M-3 equipment not only for the medical or the crew status passes, but we used it regularly three times a day. It is a very valuable piece of equipment. It came in very handy and it certainly was useful for this reason. I guess there is some reason for it for medical data. I did not understand why we could not stow it the last day. We had already checked and they said, "All right, go ahead and stow it, you do not need it." Later they said, "You have to unstow it. We want to get one more pass on you the last morning of the flight." After 14 days of flight I did not understand why we needed it, but we did it. Jim left it out, actually. You left it right over the circuit breaker panel, didn't you? Did it come out during reentry?

LOVELL — No. There was no problem.

BORMAN — There is no water problem.

LOVELL — There might have been a little air in it because we got air in the food all the time. I do not know how it got in because the food packages were evacuated. We would put the gun in and pump it up with water, and yet there was air in the food, every time you opened it up. There was probably some air in the water, but it did not bother us too much.

BORMAN — No. I thought it was a minimum amount, too. The water tasted good. It was cool. The gun, as we have already commented, was very adequate. I think it is inconceivable that we continue to have to log drinks the way we did. I think if people want to know how much water you drink, you can read them off the counter on the gun, and that is it. We went crazy logging these things by numbers and counter numbers and everything else. It is operationally unacceptable.

LOVELL — For flights that are not primarily medical all they have to have is a counter reading once or twice a day for the systems people. They could just divide it by the number of crew and come out very close to what the actual consumption per man is.

BORMAN — The food, I thought, was by and large very good. One suggestion on the food is that they try to reconfigure the meals so that Meal A is more like you would think of as a breakfast, with maybe some toast, cereal bars, and sausage patties; rather than fish, potato soup, and clam chowder for breakfast. The idea of making our day like a regular Houston day was a very, very valuable one. It would also be rather nice to have the meals correspond to the type of meal you would eat on the Earth. We ought to have a breakfast that is breakfast; and so on with lunch and dinner. One breakfast we had shrimp, sauce, peas, and I think potato soup. This is all right, but it would be more desirable to have had something like cereal cubes and sausage patties and things like that, something you are more used to.

LOVELL — The disinfectant pill crumbles. They would crumble when we got them out. The pieces would float about, and if they got in your eye they burned because it is a chlorine base pill of some sort. It happened to me once early in the flight, and it happened to Frank towards the end of the flight. We had to use the exhaust hoses to vacuum down the spacecraft to get rid of these things so we could not get them in our eyes. I think that we can probably go to something better in the future. There is a lot of room for improvement in food. It was good. It was adequate. We lasted 14 days. We could have lasted a lot longer on the food. But that does not mean there is not room for improvement.

BORMAN — The concept, as far as packaging and everything goes, is good. What is lacking now is really an adequate quality control for uniformity. If every one of the food bags had been as good as the good ones, there would have been no problem. Some of them did not have velcro on them, some of them burst, and that sort of thing.

LOVELL — You could get the soups out of the spout very easily, but trying to get tuna, salad and shrimp and sauce out of there was a real job. We should change the size of the spout according to the type of food we have. We noticed at the beginning of the flight that dry solids were especially bad.

BORMAN — Yes. Beef bites and bacon and egg bites are horrible and should be deleted from the menu.

LOVELL — GT-5 reported that the beef bites were crumbly. Every single package of beef bites that I got out was crumbs. They would float all over the place, and you had to get out your exhaust hose and gather them all up again and throw them away. If we did not have that technique, did not have the screens on the exhaust hoses, they would either go into the ECS system or float around. I could see where GT-5 got an awful lot of crumbs floating around.

BORMAN — Sleep periods. This is one of the areas where we really made a wise decision. We decided that we would sleep simultaneously on the regular Houston schedule. We did slide it back every day to correspond with the precession of the orbit. When we were scheduled for house-keeping and sleep, we would close up the windows. We found that the polaroid filters were not adequate, so, we cut up an aluminized food bag and placed it between the window and the polaroid shield. Then, it was really dark inside, it cut out the heat, and this left us with a real simulated night. As far as we were concerned, it was night time. We would get up the next day, go to work, and it kept us regular. It kept us relatively on a constant type of schedule. I thought it was very, very good.

LOVELL — We are going to have to go to that for any of the long flights, any of the lunar missions. For any of the long flights we are going to need to use a regular Houston or Cape day and not change the routine.

BORMAN — On Apollo, with three men, you probably will have to stand a watch.

LOVELL — True. But still, you are going to have to keep from getting too irregular. I thought sleeping in zero-g was very comfortable. I slept like a log that first week.

BORMAN — Yes, I had troubles sleeping. The M-1 was the culprit. It was a pain in the neck. We decided to leave it on though on the theory that if we turned it off the first thing the experimenter would say was, "Well you turned it off. It was not a valid test." Then some crew in the future would have to fly with the thing. So, we left it on for two weeks and listened to it "clank." As far as I know it didn't do any good. Maybe that's the end of the M-1. Invariably, in a state of semi-consciousness it would rouse me again. I did not like that.

LOVELL — Sleep configuration was very easy. You just clasp your hands together and hold them there. When you wake up your hands are still clasped together. There are no pressure points. You can have a book up there, go to sleep holding the book, and wake up the next morning and the book is still right there, still at the same page. It was outstanding. If mattress companies ever find out how to make a zero-g mattress they would really have a fortune.

BORMAN — Sleep period mission briefing. It followed the way we were briefed. Very good.

LOVELL — Flight controllers were very outstanding. Keeping quiet during the sleep period. As a matter of fact, they even called up Wally one time and told him to be quiet. Yes, they told him to be quiet. It was our sleep period.

BORMAN — Everything went fine.

9.0 OPERATIONAL CHECKS

9.1 APOLLO LANDMARK INVESTIGATION

BORMAN — We should first mention the weather because this casts some reflection on the whole idea of Apollo landmarks as a navigational aid. The weather was the big bugaboo in this flight as far as achieving any Apollo landmark photography. I do not know whether it was the particular targets we were trying to get or what. Invariably, there were clouds.

S65-63926

LOVELL — As can be seen from the map, anything south of 15 degrees north latitude in Africa is no good for Apollo landmarks because it is invariably cloudy. South of 15 degrees north latitude is invariably too cloudy for Apollo landmarks.

BORMAN — Right.

LOVELL — All of South America is out. As a matter of fact everything south of 15 degrees north latitude all the way across the map is no good. We found out that North Africa and Southwest United States and parts of Mexico, as previous crews stated a long time ago, (that includes Saudi Arabia, Pakistan Valley) have predominantly clear weather in the morning, but not in the afternoon.

BORMAN — The Red Sea area was one of the clear areas all the time. Even the Southwest U.S. and Mexico were clobbered most of the time on our flight.

LOVELL — Right. Weather is the big bugaboo on Apollo landmarks, or using Earth landmarks for Apollo.

BORMAN — The acquisition data was good. The pointing data was good on all the experiment updates, except for one. They missed the time on an S-5. We caught it, and did it ourselves because we knew where the area was supposed to be. The pointing data given was great throughout the flight. The updating was fine. We had no problems at all. Pointing instructions were good.

BORMAN — Concerning taking acquisition photographs for the Apollo landmarks, it was pretty darn hard to determine anything when we pitched 30 degrees down to pick up the landmarks. We could not pick out the air fields or anything. We would point the camera and take a picture, but we did not know what we were taking pictures of. We could not see it with the naked eye. I am thinking of the airfield in Brazil. We took this acquisition picture without any idea of what we were taking a picture of until we finally got to thc nadir. Then, we looked down and could see the runway. I think it was in Belam Province in Brazil. The weather was a very definite factor in photographing the Apollo landmarks, and of course, it would be a very definite factor in using the Apollo landmarks.

LOVELL — We had to use the sun angle that we had at the time we got over the target.

BORMAN — The people who called them up were taking this into consideration, because the sun angles for all the Apollo landmark attempts were good as far as photography goes. Sightings were tough on a lot of them, primarily because of clouds.

LOVELL — Apollo landmarks of interior Africa, which they gave us several times, (like islands, lakes, Leopoldville in the middle of the Congo), were very difficult because there is nothing down there but jungle and little streams and things.

BORMAN — By far the best landmarks are interfaces of beaches and water.

LOVELL — That is right. Sandy beaches with blue water. There is no doubt about that.

BORMAN — There is good contrast, and with a good map you do not need the photographs. I do not understand the value of the photographs. I do not see why they do not engage the Army Map Service to aerial photograph these areas and print the pictures according to the scale they want. It would be a relatively easy job to do and certainly would be much less expensive than taking the fuel to do it on an operational mission.

BORMAN — The designated targets were clouded over more often than not. We did take some alternate targets and pointed out some prominent features along coastlines in Africa. We have them logged and we will be able to go over them with the Apollo landmark people. Even an area like Dakar, which you would think would be a relatively clear area, we tried three times to photograph and each time it was cloudy.

LOVELL — Mostly in the afternoon.

BORMAN — The maps and Apollo landmark data package was all right. One of the problems with the maps was it is difficult to orient the map segments on the page, so that several of the targets were anywhere near the middle of the map. You might end up with a map with an Apollo landmark right on the edge of it. It is then difficult to associate the surrounding terrain with it. The big map that we had in the front of the book solved a lot of this problem. We could get the big picture from the big map and then go to the detailed one for the detailed pictures.

BORMAN — Photographs, 350 and 352, the Cairo area, you don't need a photograph of the Nile River running into the Mediterranean. That was one of the most prominent features we had. The Suez Canal, the Red Sea, and the coast of Arabia were loud and clear the whole flight. The junction of the two rivers, the white Nile and the blue Nile, were also very prominent. Again, you did not need a photograph to determine those.

LOVELL — The Red Sea and the Gulf of Eden as it goes into the Arabian Sea were very prominent.

BORMAN — Yes.

LOVELL — The 90 degree bend was very, very prominent.

BORMAN — We have a difference of opinion here, but I thought the maps were entirely adequate. Mountains were, as far as we could tell, adequately portrayed, although, we did not have any landmarks that were in the mountainous area. Most of ours for some reason were in the tropical areas and they were cloudy all the time. The cities that we saw were by and large over the United States.

LOVELL — We did not really see much in the way of cities. We saw the Australian cities at night, we could see the lights from them.

BORMAN — We saw small towns along the coast of Mexico.

LOVELL — Also South America. I wonder what the difference in altitude does to visibility of landmarks? I am sure that your visibility is going to really go down. I think that is what happened.

BORMAN — Maybe, although when we were in an orbit of 120 by 178, I could not tell the difference. One difference of course, is we were in the darkness at 178. When we were really looking we were around 120 to 140. I never got the feeling that I was going up hill or going down hill in the elliptical orbit.

LOVELL — We saw one good airport. I thought it was Ellington, and it turned out to be Houston International.

BORMAN — We also saw that one in South America very well. It stood out loud and clear, that white runway on the Apollo landmark. When we finally got over it there was a break in the clouds and at the last minute we got a picture of it.

LOVELL — Airports in general would stand out.

BORMAN — Coastlines are the answer. They are by far the best landmarks you have. I do not think islands are too good for landmarks unless they are relatively large; small islands are tough to pick up.

LOVELL — And they are usually covered by clouds.

BORMAN — Color contrast between land and water was very good, particularly along sandy beaches. We had onboard some photographs that were taken on GT-5 with the actual scene we were viewing. We were able to compare the color of these photographs. We found it then to be very, very close. One strange phenomena is that greens don't come through. The very green jungles of Brazil and Africa appeared almost a brownish-mustard color. The predominant color is blue, even at night time with a full Moon.

LOVELL — I thought the jungle even looked blue, bluish-brown.

BORMAN — Perhaps, the colors were a little deeper in the photographs than were the landmarks but photographs are very adequate presentation of colors. I think probably any variance in the photographs is due to variance in sun angle.

LOVELL — We understood the targets well enough, I don't see any problem there at all. We didn't use the maps with gloves on. You can hardly use anything with gloves on.

BORMAN — The best thing you can do with gloves is take them off and stow them as soon as possible.

LOVELL — The Apollo Landmark Book size was adequate. Any smaller size in maps gets to be too thick or too small, and you really can't read them in detail. I did like the big map of the world broken up in four sections, because that gave us a good approach to the targets. We could use that much better than we could the orbital plotboard.

BORMAN — Probably all you need is that big map cut up in four sections and then a small scale map of the specific area. We don't need the darn photographs.

LOVELL — It's the thickest thing we had to stow.

BORMAN — Spacecraft control. As far as tracking ground targets, you could always achieve the rates in PULSE mode. Fuel was nominal. PULSE was no difficulty. We tried to take everything at a 90 degree angle. We had no problems with spacecraft control to acquire and take pictures of the Apollo landmarks.

BORMAN — Certainly this factor of weather leads me to believe that the whole idea of navigating Apollo by a landmark needs to be reevaluated. It seems to me that a much more desirable feature would be a series of radar beacons placed throughout the world, similar to the ones that are used by SAC, or some type of electronic gadget not dependent upon clear weather. You cannot choose the weather you want, and in our fourteen day missions for about eleven days we had lousy weather. The number of primary targets photographed. How many did you photograph?

LOVELL — We have nine Apollo landmarks logged in the Apollo landmark sections:

Sequence 85 – clouds at nadir.
Sequence 137 – no luck, clouds overhead.
Sequence 94 – unable, because of clouds. Took a point of land nearby.
Sequence 58 – Cloudy.
Sequence 70 – Okay.
Sequence 85 – Just weather is the remark.
Sequence 108 – Clouds, resulting in poor picture.
Sequence 130 – Weather.
Sequence 97 – Weather.

Most of the landmarks they gave us, that were down below the 15 degrees north latitude ended up with weather.

BORMAN — The number of primary targets photographed was 1. Evaluation of sequence 350B and 351B, I maintain that the maps are entirely adequate. We evaluated them looking at the Cairo area and the Dead Sea, and I see no reason for the photographs. Jim has a different feeling about that.

9.2 CABIN LIGHTING SURVEY

BORMAN — We did not have. We have already talked about the lighting, glare, the dirty windows, and so forth. I felt that this was a pain in the neck, and I would like to make sure that somebody got some useful data out of that. We went around the Earth in the HORIZON SCAN mode once with the HF transmitter on, and I wonder if there is anything at all going to be learned from it.

LOVELL — I am beginning to wonder whether the HF system is worth the cost to put it into orbit.

BORMAN — You might be better off with another UHF set.

LOVELL — That is right. Go to something that is compatible, like single side band.

BORMAN — HF was adequate when UHF was adequate. As far as any over-the-horizon transmission with HF, there was not any.

LOVELL — It was very, very poor.

BORMAN — The reception on HF during the flight was very poor.

LOVELL — Right.

10.0 VISUAL SIGHTINGS

10.1 COUNTDOWN

10.2 POWERED FLIGHT

65-HC-1123

BORMAN — We did not have any visual sightings. No wasps, bees, bugs, or anything in the spacecraft.

BORMAN — I did not look out at lift-off.

LOVELL — I think we went through clouds, didn't we? We went through an overcast.

FCSD REP — It looked like Frank steered it right through a hole in the overcast.

LOVELL — Could anyone see us after we went through the overcast, or was that the end of it?

FCSD REP — No, the hole was big enough so that you were in sight.

LOVELL — Could you see staging?

FCSD REP — No, you disappeared from sight before staging.

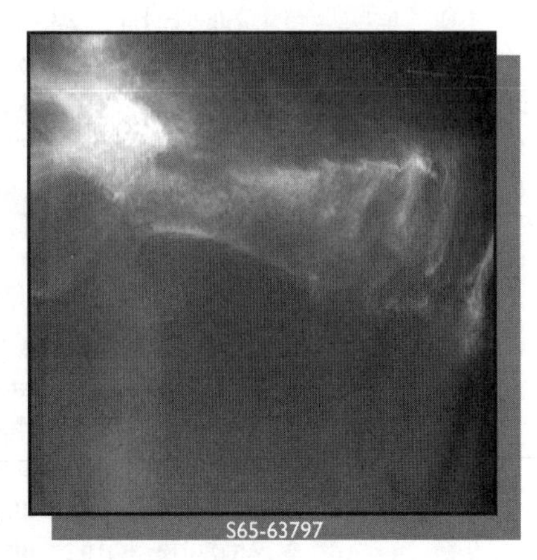

S65-63797

BORMAN — At BECO the spacecraft was enveloped in a yellow flame. BECO and staging are so close that you cannot tell the difference. Engine 2 ignition was very smooth, no noise, no banging. The horizon came into view just like the simulator. Very, very adequate presentation. At SECO I did not notice a thing, just a stop and a cessation of noise. I did not see any flames, debris or anything. Although you did at Fairing Jettison and Spacecraft Separation.

LOVELL — On Fairing Jettison, I saw debris fly, especially during the turnaround.

10.3 ORBITAL FLIGHT

BORMAN — As for man-made objects sighted, we took sightings of the booster. We had no difficulty in sighting the booster at turnaround and in station keeping with it. We maintained the flashing light in sight throughout the first orbit, we saw them on the second orbit, and the third orbit during the night phase.

LOVELL — But you could not tell distance by the flashing lights.

BORMAN — No. As a matter of fact, I think it would be impossible to tell distance at night unless you have some sort of illumination.

LOVELL — Yes, if the lights are illuminating the vehicle, by spreading it over the vehicle, maybe you could tell. You cannot tell the way they are now. We picked up Spacecraft 6 on the final rendezvous phase, I guess at about five miles. Is that right?

BORMAN — Right. Again, this is entirely dependent upon where the sun is. Spacecraft 6 was able to see us a long time before we could see them, because we were in a position where the sun was reflecting off our adapter, and they were in a position so that the sun was not reflecting off their adapter. It is a strange phenomenon, but it is entirely dependent upon the sun. Once we acquired Spacecraft 6, it certainly was no problem of maintaining a visual on it. It was no different than it is on the ground in the daytime.

LOVELL — I agree.

BORMAN — We sighted two satellites. We saw the Minuteman reentry at exactly the right time and exactly the right place. It looked like a meteorite.

LOVELL — Very brilliant. It broke up toward the end.

BORMAN — Very brilliant, it broke up toward the end, and there were several different pieces flying around. I was surprised at the speed at which it went in. I was also surprised at the control authority required to track it. We did get on it though before it broke up and went out. I hope they got good data on it.

LOVELL — We saw, what we think, were two satellites moving.

BORMAN — Yes, we took IR readings on one of them. It was a definite satellite in a polar orbit. We took photographs of two satellites in the polar orbits. One of them was below us, crossing from left to right, and the other one was above us, crossing left to right. We picked them up first in the handle of the Big Dipper. It was definitely a satellite, and we tracked it on the IR and we used the recorder. It will be interesting to see if they got any data of it. We were never close enough to see any satellites and to pick them out as far as definite features.

LOVELL — They were just points of light.

BORMAN — Just light points. There certainly was not any problem viewing the ground. As we have already mentioned over the radio, we were able to see the Astrodome. The horizontal markings, or the east-west markings, on the S-8 / D-13 pattern stood out loud and clear when we acquired it. It is very akin to flying an airplane at high altitude. Roads stood out well, rivers stood out well, and beaches particularly well. Towns were a little more difficult to determine, but when you had them you could easily see the layout and apparent size. Also, the atmosphere made a vast difference. If there was any sort of haze or cloud, it definitely cut down your vision. When we finally got over the Himalayas in daylight, it was easy to see why it was so clear.

LOVELL — Haze does affect sightings from space as well as it does from airplanes. It is very similar. If the area is pretty hazy, it is going to decrease the contrast of the ground, and makes viewing ground objects very difficult. Where you have excellent viewing is where the air is clear. It is the same way with airplanes. But, I think you have a better chance to look below than you do in an airplane. There is a better chance of seeing something, but you still have the haze problem.

BORMAN — When we first got up and had a full Moon, we did not see many more stars than you see on an ordinary flight at thirty or forty thousand feet. However, during the latter part of the flight with a quarter Moon, we found that there were more stars visible. We have a drawing here of the number of stars we could see in the Pleiades, which has been recorded.

LOVELL — The big thing to remember here is the fact that the Moon has a big effect on the number of stars you see and which ones you can use for celestial navigation.

BORMAN — Right.

LOVELL — If the Moon is full, it is the predominant feature in the sky, and any stars near the Moon at the time are washed out; and they are very hard to use for sextant sightings.

BORMAN — Right. Same way with the Milky Way. It is not very apparent when the Moon is full.

LOVELL — In reverse though, when the Moon is full the Earth is a very good reference.

BORMAN — That's right. It's a blue reference with white clouds, very visible. Cloud coverage, there was a lot of it. The first three days were relatively clear and from then on there were lots and lots of clouds. About the only area that was consistently clear was the Sahara and the Red Sea area. The States were cloudy, as was Mexico, South America, Southern Africa, and a great deal of the ocean was cloudy.

LOVELL — The entire area of the Pacific Ocean was cloudy.

BORMAN — Again we can reiterate that it would be pretty tough to count on being able to observe specific landmarks at any given time. The horizons were interesting. In the daytime I never noticed much of a difference between the horizon. We've drawn pictures of them with little notes beside them that have been carried back, and which will be published. We noticed the difference in the horizon and the air glow layer between a full Moon and a no Moon. When we had a full Moon there was a definite Earth layer, then almost a normal blackness, and then a definite band of air glow. When we had no Moon or a quarter Moon, the area between the Earth and the outer air glow was more milky.

LOVELL — That's right. For sextant sightings the best thing to use, during a full Moon, was the bottom part of the air glow or the Earth's horizon, which was probably the upper air horizon. And with no Moon, the best sextant sighting spot to use was the upper air glow, because it was the sharpest of the two. The area between the upper air glow area and the Earth's horizon was too milky during a no Moon to get a good sharp contrast (you could see stars through it), but you couldn't find out exactly where the horizon was. So it varied between full Moon and no Moon.

BORMAN — Very drastically too. Also, your ability to pick out the Earth, and determine yaw from Earth's passage with no Moon, was zilch. With the full Moon it was just as good as it was in the daylight.

LOVELL — The Moon is very bright up there, very bright.

BORMAN — Both attitude and translation thrusters firing were apparent. The aft firing thrusters were not apparent. But the forward thrusters were apparent at night and so were the translational, up, down, sideways. The attitude thrusters were also apparent with puffs, sort of like subdued flashbulbs appearing out to the side of the window. They were just little flashes of light in the background, and we couldn't determine anything from them.

LOVELL — We couldn't determine anything.

BORMAN — But you could on Spacecraft 6.

LOVELL — On Spacecraft 6 the thruster firings looked as though they went out about forty feet against a dark background. You could see this light haze going out about forty feet. It wasn't a bell shaped pattern, as we thought it was going to be, but it was like a garden hose, straight out. At a distance we could not see the bright flash of ignition in the chamber itself. All we could see was the haze going out. But as we got in closer, we could start picking up the bright flash, mainly because of attitude I suspect, as we were getting closer to see it. It had no effect on our spacecraft at all. They actually fired about twenty feet away with their forward firing thrusters to us, and departed.

BORMAN — They got in back behind us in the adapter and fired their forward firing thrusters and made the curtain jiggle. I don't know how far away they were when they did this. I hope they can tell us, and give you some idea of the effect that these thrusters might have on the EPA Mission. It's just like firing a rocket at night. I guess that's the way we expected it to look. That is exactly what you are doing, firing a rocket at night.

10.4 REENTRY

BORMAN — We retrofired in the darkest, most dismal part of the night sky, and we didn't see anything. We didn't see Adapter Separation. We didn't see any of the retros fire. We didn't see Retro Package jettison. At reentry we didn't have a lighted horizon at 400,000 feet. We didn't have a lighted horizon until we were below 400,000 feet. This was not particularly to my liking; although we picked it up about 330,000 feet. After then, it was all right.

LOVELL — We saw ionization.

BORMAN — Yes, we did.

LOVELL — The whole window started to be just a glow outside. Then it got into the rings of this ionization layer. It started out to be just glowing.

BORMAN — When we went subsonic it looked like there was a bunch of steam and flame that engulfed the nose section. I think this may have been when the spacecraft finally went subsonic, right before drogue deploy. You could see the spacecraft oscillations, and you could reference them with the drogue, which was relatively steady. You could also reference them with the sky. You didn't have any stars or anything, but you could just pick out the motions across the sky. At times it was pretty violent. It would be interesting to see the magnitude of the oscillations and the rates.

BORMAN — The drogue deploy was very evident. We saw the drogue chute. We saw the R and R Can come off, and then I saw it following us down in the parachute, off to one side. Main chute deploy was very nice.

LOVELL — Main chute deploy, very enjoyable to watch.

BORMAN — During landing and recovery we had the S2F in sight before we ever hit the water.

11.0 EXPERIMENTS

11.1 CELESTIAL, SPACE, AND TERRESTRIAL RADIOMETRY (D-4 / 7)

BORMAN — I think we have more valuable data on D-4 / D-7 than all the others put together.

LOVELL — Right.

BORMAN — Updating techniques and communications procedures were excellent. Everything called up on D-4 / D-7 was right on the money, including the time and pointing locations for the reentry vehicle.

LOVELL — They did an outstanding job of giving us the right information onboard.

BORMAN — We were perfectly trained in equipment setup and usage. Jim didn't have a question about it. I think there was no problem at all.

LOVELL — We got everything going. I hope that we got some good data.

65-HC-1046

BORMAN — Cooled Spectrometer checks were nominal. I don't know what else to say, as far as we know, everything was perfect, it's all logged, they have the records, they have the TM.

LOVELL — We saw the needle go down when we jettisoned the fairing cover after the fifteen second timer delay, after we did our separation maneuver. I saw the cover go by when we jettisoned.

BORMAN — Did you?

LOVELL — It is right above the right hand window.

BORMAN — I did not see that. The booster measurements were no problem at all. It was just like the way the simulator worked. It was just fine. Very good. We had no problems making the reticule measurements. They were very close to zero on both the IR and the spectrometer. Powerdown procedures. Did you have any?

LOVELL — No. Booster background measurements were no problem at all.

BORMAN — Okay. The Milky Way measurement without the cooled sensor was performed exactly on time. We had no problems with the equipment. No problems with any of this. The void was very easy to pick out and to take the readings. The Zodiacal light test was deleted in real time. We substituted Sirius for the Zodiacal light.

BORMAN — The star measurements were done with no problem. Night land measurements are noted on the tape and in the log book. We did have some problem with these because the night land was covered with clouds. So we were trying to skew the spacecraft along and avoid clouds. I'm afraid that we may not have always missed the clouds. The same situation existed with the night water measurements. We had very good luck on the Polaris.

We had good luck on the Polaris and good luck on the booster. We should have had pretty good luck on Gemini 6. We were able to track it. We had good luck on the sled.

LOVELL — We hope we had good luck on the sled.

BORMAN — They called us up from the ground and said they had a thousand percent on that.

LOVELL — You could track the Polaris launch in PULSE.

BORMAN — You could track everything in PULSE except the reentry vehicle. The reentry vehicle was pretty fast and you had to go to DIRECT.

LOVELL — I have to give credit to the personnel, the experimenter, Brentnall.

BORMAN — He hustled and he did his work. He made sure that we knew what we were doing. It was well done.

LOVELL — This is the first missile that was ever launched from a submarine, their ORI, I guess. To launch at the exact time so that we would be in the exact position, I thought was an amazing piece of coordination among all people involved.

BORMAN — I just hope they got all the data.

BORMAN — We always had the ACQ off when we were transmitting. The only possible problems that we could have had with this experiment was when we lost our tape recorder. We could not put some of the data on tape. We had to always take the data over a receiving station.

LOVELL — Or use the recorder.

BORMAN — Or use a recorder.

LOVELL — One problem which we did have and which I was not briefed on was the fact that I inadvertently left the gear on one time and it gave a nine second pulse on the main ammeter system. We could not figure out what this glitch was coming in on the Main Bus.

BORMAN — This was right after we lost the thrusters. We thought there might have been something in the ACME bias power or something in the inverter that was breaking down and putting out a glitch. We called the ground to tell them we found it and they said that it was exactly what the ground had thought it was.

BORMAN — Cloud illumination lightning; we got some good lightning shots for them. Day land measurements, we got. Ascension Calibration was done twice.

LOVELL — Listen, that was done twice?

BORMAN — Yes.

LOVELL — Was this the NADIR to 35 degrees West?

BORMAN — Yes.

LOVELL — Of course we never saw Ascension so I did not . . .

BORMAN — Yeah, we never saw Ascension but it was the location of the area where we wanted to find.

LOVELL — I think we did that once, or did we do 19 once?

BORMAN — Yes, it is the same thing. Cumulus clouds. We got some cloud measurements. Star measurements, the only thing we got there was the lightning. Celestial measurement. We got that. The stars, we did not get the Zodiacal light. We got the stars, the Milky Way, the void. 23 Moon measurements. We got the Moon. We got all the missile measurements. I do not think it is necessary for us to record it off our log book.

LOVELL — We missed the Titan launch.

BORMAN — Yes, we did not get the liftoff because of clouds. But we picked them up when we picked up their contrails.

LOVELL — Yes.

BORMAN — We also threw in an extra satellite. It will be interesting to see if they got any data out of that. The Sun measurement was done. That was a real interesting one, because in order to get the Sun, we had to put up the polaroid lights, and wear sun glasses, and then point at it.

LOVELL — Hot Earth Measurements: we had that big fire in North Africa that was there all the time. That was a target of opportunity. We got Hot Earth Measurements of this fire in North Africa.

BORMAN — The tape recorder was used properly. Voice recorder usage was adequate.

Flight control procedure: the only time we had to do anything at all different was to go to DIRECT to pick up that reentry vehicle. The rest of the time it was a very well organized, well briefed, well trained, and I thought a well-run experiment.

LOVELL — That is right. I thought it was quite a bit of data and work to do in that experiment.

BORMAN — As a matter of fact, I think the equipment on these represented about 3½ million dollars. So it was a pretty hefty piece of equipment.

11.2 STAR OCCULTATION MEASUREMENT (D-5)

BORMAN — Now we go from the sublime to the ridiculous. The star occultation measurement D-5. The equipment never worked and we fooled around with it for hours trying to make it work. I am sure somebody is going to sit downstairs now and ground check it okay, but, it did not work.

LOVELL — We did everything and we traded it back and forth to make sure that both of us saw the same results.

BORMAN — We had been briefed on this, we used it in the set up at St. Louis where we had seen how it changed. We had had adequate briefings. We had accurate presentations on it.

LOVELL — Primarily you could not calibrate the equipment, the reticule would not change from red to green.

BORMAN — We made a couple of measurements where we just used the calibration knob at a certain setting and let it alone. But this equipment was inoperative. I thought it was a little strange, too, when they called up and said, "We have just found out that this equipment is sensitive to RF interference. Turn off all your transmitters and see if this affects it." I think we ought to know before we go charging into the 3rd or 4th day of flight that the equipment is sensitive to RF interference. It seems to me that perhaps there was not a very good Qual. test done on this piece of gear. I have nothing else to say about the D-5 except the equipment did not work.

LOVELL — Well, star acquisition and identification, which has nothing to do with the equipment, was outstanding with the chart. Actually, I would like to put in a little pitch here, for the simulator, because the visual part of the simulator was a big help. It gave us a good idea of what we could expect in flight. And the Planetarium's work was very valuable in conjunction with the simulator. I think that we did not waste time with any of our star recognition stuff. They did come in handy. As a matter of fact, if you really get to know that celestial sphere, you can really get around without the Earth.

11.3
SIMPLE
NAVIGATION
D-9

BORMAN — Jim used this test exclusively. He was trained at Ames. But we actually used the sextant down here before the flight many times with the experimenters. We used it in the simulator.

LOVELL — Let me give some comments on this sextant. There are a lot of comments on the tape. Number one, the D-9 sextant alignment was fine, the stop watch was an excellent idea. The lighting on the sextant angle reading was too bright. It was a white bright light. Every time you read the sextant out you lost your dark adaptation because we were looking at this white light. And all it had to be, since the spacecraft is dark anyway if you are looking at stars, is red, or dimmer light to read the angle out. The reticule was fine during the first part of the flight, and then for some reason it became a double image. I found out that I never used the reticule during the full Moon or even at half Moon I could see the dark lines without having to light up the reticule. One of the big things about the sextant, the mechanical aspects, was the fact that the light was not split 50-50 between the upper prism and the lower telescope. It was a 20-80 split and therefore one image would be bright and the other one dim. This was fine if you went from the start of the horizon where you would have a bright star and a dim horizon. You had a control, but if you went from, say, two stars to measure the angle between two stars to check the alignment of a sextant, (because the angles are already known) one star would be bright and the other star real dim if you got them within a certain angle of each other. But as soon as you got that other dim star, then the bright star, you could not tell when they were overlapping.

LOVELL — You had to have two stars of the same magnitude in lock. So, I had practiced and trained in the one with the 50-50 split which I thought was a lot better than this 20-80 which we finally ended up with. And this also is true when you took star to Moon limb shots because they had it arranged that most of the light came from the Moon. Eighty percent of the light passage was from the Moon and they had filters to cut that down. And 20 percent came from the star. Well, the filters were not dark enough for the Moon, and as soon as you got those stars somewhere near the limb they just faded out of view. You could not see it so you had a hard time getting star to Moon limb shots.

LOVELL — We noticed one big thing in navigation with the sextant. With a full Moon, you are going to have a hard time using the stars under the Moon. It just blanks out the light. It is just too bright, it dims the stars and ruins your dark adaptation. On no Moon nights or quarter Moon nights, most of the stars come out and the navigation is a lot better. We found out also on full Moon nights as we said before, that we'd use the bottom Earth's horizon, which is probably an upper air horizon and had the sharpest line of identification for angles.

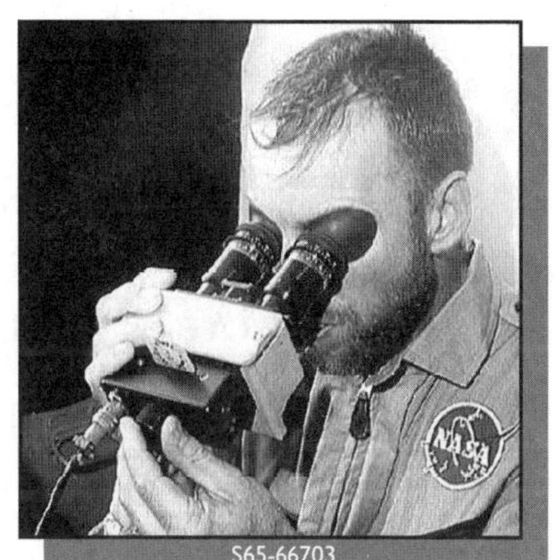

S65-66703

LOVELL — On no Moon nights, the whole air glow layer seemed to be of the same consistency. And the sharpest part was the upper air glow layer where we could bring stars down to it.

BORMAN — Did you find it difficult at times to get the stars acquired at the angle?

LOVELL — Right. We found out, due to the window design, that if the stars were not lined vertically with one another, if we were taking star to star shots or star to Moon shots, it was hard to get the sextant lined up. The best way to hold the sextant was the way it was developed originally: just up and down. If you had to move it around here to get a star over here, or horizontal or angle shots, you had to move the spacecraft around to get the window in line, to get all the view. Now we could shoot angles up to about I think 45 degrees, if we had the right window alignment and had bright enough stars. But anything above that and most of the time around 35 to 40 degrees, one of the two, either the upper prism or the lower telescope would get in the way of the rim of the window and get lost to view. Then we couldn't do anything with it.

BORMAN — I remember there were a couple of times when you picked up the wrong star too.

LOVELL — Yes.

BORMAN — Even though you know the star very well, using the system of estimating the angle in ahead of time, you leave your self open to picking the wrong stars.

S66-17442

LOVELL — Yes, one of the big things about the sextant is you have to be sure you can identify the star when you are looking inside the sextant after you get the thing down because you might get the wrong star. That is why you should use stars in sextant navigation that have identifying features in the field of view of the sextant. For instance Capella was a very good one because it has the 3 little stars around it, I guess there were 2 stars, I forget.

BORMAN — Three stars.

LOVELL — Three stars, but anyway you could tell Capella when you looked through the sextant. But, a star that was all out by itself, even though it was bright, you really could not tell whether that was a star or not until you went out and looked again, and that was sort of difficult.

BORMAN — We did not have any radar but I know darn well, if I was out trying to effect a rendezvous, I would much rather have an operating radar with range and range rate than I would that sextant.

LOVELL — Well the thing is: there are a lot of factors that affect this thing: window glare, the fact that we had to turn on white light every time we got a reading, because we had to read the digital clock to get an exact time of the reading, the reflection from the Moon, all that stuff complicates visual sightings.

BORMAN — How about picking out a proper horizon? You even had trouble doing that. What horizon were we going to use?

LOVELL — That is right. We don't know what horizon we are going to use.

BORMAN — The horizon is dependent upon the Moon.

LOVELL — So sextant sightings are going to be very difficult.

BORMAN — And the accuracy of this instrument, as advertised, is phenomenal. But the accuracy to what? If we do not know what horizon we are measuring to. Did you mention the fact that the green filter cuts out everything?

LOVELL — Well that is right. Yes, the green and blue filters, one of the modes as a matter of fact, was to use the green filter for the horizon. Unfortunately you stick the green filter in there and then look for the horizon and it is gone.

BORMAN — Everything is gone. There is nothing. All you see are the stars up above, you do not see any horizon.

LOVELL — So the filter idea was useless as far as the sextant goes.

BORMAN — Also the blue filter, you stayed dark adapted and then tried the blue filter in daylight as was requested.

LOVELL — That was a little scientific experiment which we were asked to do: to see if we could see the blue horizon line, blue air glow in the daytime by keeping dark adapted and using the blue filter.

BORMAN — The results were negative.

LOVELL — That is right.

BORMAN — Anything else?

LOVELL — I think we hit everything on the sextant.

BORMAN — We had no equipment malfunctions with the sextant other than the fact that you got double vision.

LOVELL — And the sextant was not as bulky to operate on the spacecraft as it is on land. It is a pretty bulky piece of equipment, but it is not bad.

11.4 VISUAL ACUITY AND ASTRONAUT VISIBILITY AND VISION TEST (M-9)

BORMAN — Updating techniques and communications procedures: This was an area that was outstanding. It seems redundant to comment about it every time, but it is true. Equipment set up and usage for the vision test and the M-9 experiment, no problem. We tried this in the simulator and we tried it in flight and it was the same.

LOVELL — The photometer left a little bit to be desired.

BORMAN — Oh, I think we are talking not just about the tests that we did every morning. The vision test and the other one.

BORMAN — The M-9 vision test also caused no problem. As a matter of fact we did a little interesting experiment of our own. We used the brace, the head brace and the bite board every time but once, and then we compared the results, with and without using the head brace or bite boards, the last day. We noted this in the log book. There was no difference in the outcome, so I have begun to wonder if it is necessary. One of the great points of interest was trying to observe this ground observation at Laredo and we picked it up 3 times, I believe. Acquisition is very difficult, because of the poor terrain features. I never did really see the smoke pots, but we did see the dashes that they had in the block. We had a real good pass at it and we had it acquired excellently and I called off 3 numbers and they are listed in the log book, I think there was a 3-1-3 or a 1-3-1.

BORMAN — I would be interested to know if I was right on any of these because you could see them. I should have been able to see them on that one. 1-3-3, I guess it was we called off. No, 3-1-3 ground observation.

LOVELL — I had one pass where I saw it and by the time I got to see the squares or the rectangles we were already by it. I guess from what I could see of the thing, I think it was a 2-3-4, that was all I had of it. Basically, we had a picture from GT-5.

BORMAN — We had pictures from GT-5 with us in the cockpit to help us acquire it.

LOVELL — We could see the red ground with no strain at all. We knew where to look and it was just difficult to find.

BORMAN — You might mention that if they had gone to Yuma, if we had not had to change our launch trajectory, it would have been great. You could pick out the area of the Salton Sea there even from where we were, 3 or 4 hundred miles South, it stood out like a sore thumb.

LOVELL — I do think though that for the amount of time that you are over the target, the amount of squares they were expecting you to read are too much.

BORMAN — Yes.

LOVELL — Because you just cannot read that much. You have to acquire it, examine it, and then by that time you are going over it.

BORMAN — The window measurements. We did this; it was called to do it twice, but we did it three times because of the fact that the tape recorder was shot and they had to get real time TM. They did not get good TM on one pass, so we did it again at an opportune time. Unfortunately, the photometer had not had a chance to warm up for 10 minutes. We were at such a low fuel state then that it was either then or never, so we did it, and I hope they got the data. Now this window measurement business was always a little nebulous to me. I am not sure exactly what they were going to get out of it. But it seems like they are scratching.

LOVELL — I do not understand the procedures too well. I mean I went through the procedures, followed them exactly, but the photometer left me out to lunch. We tried to get the dope, but I do not see what they are going to get out of it.

BORMAN — We did not have much confidence in that. We did it exactly the way it was practiced. We did practice it in the simulator and it is recorded on voice tape. They came back though and said that TM did not make much sense to them, and I think that is probably right. Voice tape recorder usage, by and large this was another experiment that was well handled, preflight and during flight. We admit that we spent a lot of time on this one and with all the visual acuity data they got ahead of time, and were briefed by the principal experimenters many, many times. I thought that it was handled as well as it could be handled. The site in Australia. We were over Australia in the daylight – of course the site was not manned because of the few passes we'd have to have. When we were over Australia we didn't have attitude control.

LOVELL — We might mention one thing. I think with the long flight like we had, if we had the fuel we could have become more skilled in observing the site and probably have gotten better results.

BORMAN — No, we could not have on this one, because after the 3rd day the clouds were the factor the whole time. We had only 3 days to observe that, and everything else was cloudy.

LOVELL — Okay, it was cloudy, but I suspect that the more passes over the thing, you would get better. But we just did not have the fuel and the weather was bad.

11.5 SYNOPTIC TERRAIN (S-5) AND WEATHER (S-6) PHOTOGRAPHY

BORMAN — Synoptic terrain and weather photography. On unusual or significant subject matter. Well, it turns out we did not have any unusual or significant subject matter as far as I know. We took the sequences and we took a lot of it as general photography that was not even called up. Most of the time, it had to be done in drifting flight. We were sorry about this, but there was not the fuel available. The two sequences that were called up (where we were allocated the fuel from the ground) we took them. I hope they were well done. We did get some pictures over Brazil that have never been gotten before. They were not called up. We took them with the IR color shifted film. We got some pictures of Mexico that were not called up.

LOVELL — We got a good sequence over North Africa.

BORMAN — We got several good sequences over North Africa.

LOVELL — But I am sure that they have been used before.

BORMAN — We also took a sequence along the southern coast from New Orleans through Florida using color shifted IR to evaluate the film more than anything else. I hope this comes in handy. We got some shots of the island chains off Florida, to see if they can determine any effect from the hurricane that went through, Betsy.

LOVELL — As a matter of fact, we got a strip sequence of Cuba too.

BORMAN — All these are targets of opportunity done without attitude control, but they were all targets that were listed before flight. We got the mouth of the Amazon, and I think that is the first time that has been photographed.

Now on the S-6 the weather photography we tried in the last pass to take a successive revolution picture of weather development. We got one good shot over the Andes. We also got some good shots of wave clouds over the Andes during the latter part of the flight. We looked for hurricane or tropical storm Alice in the Indian Ocean but we never did see it.

LOVELL — We had an S-6 weather to get that too, but we never did see it.

BORMAN — Air to ground relay of data was good. Voice recorder usage, this is one area where we are not redundant. If we had time to record all the photographs we put in the log book and not on the voice recorder in an effort to save tape. Everything that was done, though, was logged either S-5, S-6, or in general photography. Now, another item that is going to degrade this was the scum on your window.

LOVELL — Yes. The window is an area which we are going to have to work on.

BORMAN — As it got worse and worse, we shifted so that I was taking the pictures, but I had scum on my window also. The major problem though, here was just lack of fuel. We could not orient the spacecraft to control it the way we wanted to.

65-HC-1171

11.6 PROTON ELECTRON SPECTROMETER, TRI-AXIS FLUX-GATE MAGNETOMETER

BORMAN — We turned the switch on and left it on. I think we should have reams and reams of data, although unfortunately with the tape recorder being shot, I do not know what good it is going to do.

LOVELL — Well, they got real time telemetry over the station.

BORMAN — Right.

LOVELL — After seeing GT-6's film of our back area, the straps hanging off our spacecraft looked like they were all intertwining in that boom area. It could probably foul up that boom somewhat in future flights. I mean in a future flight if we still had that same problem it might foul up that boom and tear off that wire. There is a wire that goes out to the end of that thing.

BORMAN — All the experiments we talked about so far were well presented beforehand. We knew exactly what we were going to do, and the only reason we weren't able to do it was 1) weather, 2) lack of fuel, and 3) in the case of the D-5, the equipment broke down. I think we got a substantial amount of S-5 and S-6 done. As a matter of fact we got everything done that was called up and we got a lot more done just in general photography.

11.7 OPTICAL COMMUNICATIONS (MSC-4)

BORMAN — Optical communications, MSC-4. Okay, acquisition of the ground target. I have never been better locked in on White Sands in my life than we were, when we went by there, and saw two blinks of the laser. We had the photograph in front of us marking where the laser was located. We saw Holoman, we saw White Sands but we never did see the beam except the short flashes.

LOVELL — The magnitude of this laser is a lot less than either Frank or I suspected after our briefings by the experimenters and by actually looking at one at Houston. Obviously this laser at Houston was a lot shorter and can be very much magnified.

BORMAN — Tracking was no problem. We acquired one over Hawaii; we were able to track it easily. The problem is acquisition of the light.

LOVELL — The light beam is a lot smaller point than we suspected.

BORMAN — And then again the big bugaboo in this whole experiment was weather. Ascension never got up, we never got ground equipment at Ascension. We never got really good passes at White Sands because we were always about 300 miles South. And Hawaii was available to us on only 2 passes. We acquired the laser on both passes, but were not able to get the hand-held laser on the beam on the ground.

BORMAN — Okay, how about the field of view?

LOVELL — The field of view. Well, let me talk about the equipment. Basically the equipment was fine, except for the telescope: No. 1) it did not have as much light gathering as was required for this particular equipment. It should have a larger magnification. 2) It had a green filter over it which was supposed to enhance picking up the beam, but all it did was fade out everything. You could not see terrain features and you needed to see terrain feature to find out where that beam was located.

BORMAN — Yes, to know where the beam was.

LOVELL — You could look out and . . . we had photographs of the area where the beam should be coming from. And we could acquire it by our naked eye by just looking out and seeing it. But, as soon as you went to the telescope of the laser, everything faded out to this green color. Unless you just happened to pick up the beam you're lost and I never was able to pick up the beam through the telescope. The filter was just too strong. I much prefer to have a clear view. So I could pick out the terrain features. The reticule of course . . .

BORMAN — Of course, we knew that was shot before we took off.

LOVELL — We never had a good check of this thing though, at night, and I wish we had had a good night pass over something, but we never did. Because I feel that we probably could have probably picked up the beam a lot better at night.

Protective glasses. To tell you the truth the one time where I was really trying to get onto the Hawaii pass I took them off. I did not use them. And they were in the way, they were cumbersome and together with the telescope, it was just too hard to find.

BORMAN — I used mine, but I did not put them on until right before Jim was supposed to transmit.

LOVELL — Of course, I had them, I had them on and I transmitted and found out that I could not see anything so I just pulled them off and started transmitting by going back . . . because as I understand it the operator really does not need them if he is right next to it.

BORMAN — The big word on this one as far as the reticule goes, we should go back to that is for the day time you do not want that green in there. At night time you are going to need a lighted reticule. You cannot see the reticule at night.

LOVELL — That is right. So you need a better acquisition device on the laser, a better, bigger telescope with more light gathering and clarity of the terrain of the target. I would imagine PULSE mode would be great for tracking the thing.

BORMAN — Yes. I tracked one at Hawaii very easily. No problems. The tracking is a nominal task, there is nothing to it. But the other business of acquiring is very difficult, particularly with the reticule set.

11.8
LANDMARK CONTRAST

BORMAN — We did one of these with the D-5 photometer. We knew it was bad, but we put the calibration needle in the full "up" position, told the experimenter that we were doing this and took the data. Now, they will have to check it to see how it worked.

LOVELL — We did just one.

BORMAN — We had an equipment malfunction. Right. No sense talking about it anymore. Spacecraft control was no problem. We accomplished all the mechanics of doing it, but unfortunately the equipment we knew was malfunctioning before we did it. Right?

LOVELL — Right you are.

11.9
CARDIO-VASCULAR REFLEX CONDITIONS (M-1)

LOVELL — Got that started at, I think it is 3 hours and 8 minutes, or something like that. At 2:39, I turned on the M-1 experiment.

BORMAN — Okay. Procedures and operational problems. We had the hose coming all the way across your whole lap to get into your right leg when it should have been put in the left leg.

LOVELL — That is right.

BORMAN — Periods of operation. The thing was operating continuously for 14 days. And I might point out it is a pain in the neck. Because of the clanking noise. Even though it seemed you would get used to it; many times just as we were about to drop off to sleep that thing would clank, and wake you up. The only reason we did not turn it off at sleep period, was that we did not want to have the people say, "Well, you did not run it the whole time so the experiment was not valid." We left it on and put up with it in the hope that we could get rid of it once and for all.

LOVELL — Obviously, it did not work.

BORMAN — So far as we know it did not make any difference. And it did not seem to make any difference inflight or post flight.

LOVELL — It was a waste of time.

11.10
IN-FLIGHT EXERCISER (M-3)

BORMAN — It is a valuable piece of equipment. We used it not only for the crew status passes, we exercised with it 3 times a day and I thought it was very worthwhile. It is a simple piece of gear but it is a good device. Right?

LOVELL — Although, I imagine we can improve on exercise equipment in the future spacecraft.

BORMAN — Perhaps, the biggest deterioration that we noted in the muscles was in our legs, and that exerciser was – I do not know how you can improve on that much.

11.11 IN-FLIGHT PHONO-CARDIOGRAM (M-4) AND IN-FLIGHT SLEEP ANALYSIS (M-8)

BORMAN — Well, the equipment problems with M-8 were that the thing is operationally incongruous. You cannot have those wires on your head and stringing down the back of your neck, and not expect to catch them on something in a small spacecraft. Now, we found that we couldn't keep the helmet on. I kept it on for two days, but my head became extremely hot, and I was uncomfortable. So when I took it off I ripped all four of the leads off. And I think that the whole thing is extraneous anyway. I felt after 14 days I was perfectly capable to judge my own condition, and my own awareness, and my own degree of alertness. I did not need that bunch of wires hanging on my head to tell me or to tell somebody on the ground how awake or how asleep I was.

LOVELL — Inflight phonocardiogram, I have no comments because it was stuck on me at prelaunch, and it stayed on me during the entire flight.

BORMAN — Well, did it bother you? Did you get any sensor problems?

LOVELL — No.

11.12 BIOASSAY BODY FLUIDS (M-5) AND CALCIUM BALANCE STUDY (M-7)

BORMAN — Urine samplings bags, well, actually the bags were pretty good. We had one break around the head. Marking was all right and stowage was fine. I might point out that the tracer accumulator we changed with no problem. The mixing bag was all right, but they left out the most important thing here, the condom device that we used to urinate into is unsatisfactory. We ended up with urine all over ourselves every time we tried to use it. It was sort of happenstance. If you lucked out, you didn't get a leak. But based on the experience that we had up there, I would think that the way to go is a simple overboard device where you vent right to the atmosphere, and urinate into a tube. Essentially the same thing you do in C-47.

BORMAN — We carried the flow meter along and the Delta P across the flow meter and the filter can be incorporated with it. I am sure this is enough to allow you just to urinate right into a relief tube right overboard. I do not see any reason to go through this stage where you have urine all over yourself 4 or 5 times a day. A very unacceptable device. Do you agree?

LOVELL — Right. I think we can improve on the urine dump system tremendously.

BORMAN — I think they ought to start looking into just a dump right overboard. You do not use a rubber condom or anything.

LOVELL — Because that thing puts back pressure which is really sort of dangerous I think, and besides that, it is uncomfortable. And when you have back pressure . . .

BORMAN — Yeah. When you have a big enough receiver and you urinate a stream into a vacuum, it is going to go right on overboard. I do not understand why . . . okay. The defecation bags were fine. And we really used them this time. The defecation bags were really put to use in this flight. I think we used a total of 15 of them.

LOVELL — Yes.

BORMAN — They were fine. I have no comments on those. It seems to be the best solution you can have to the situation.

LOVELL — For this particular type spacecraft.

BORMAN — For this particular type . . . For the Gemini Spacecraft. I do not know what other approach you could have to them. I thought their marking was easily done. The finger hole I never used that at all. One of the big assets, we might as well put it in here, was the fact that the food that we ate caused well formed bowels, so we did not have any loose stools or any of that problem. I never had any occasion to use that finger device in there at all. Any other comments?

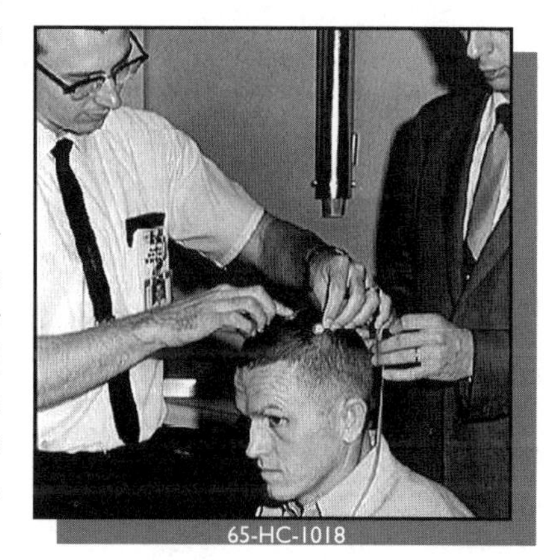
65-HC-1018

LOVELL — No.

BORMAN — The water intake, the water counter number. The gun was satisfactory. The recording that goes with it is completely unsatisfactory. I think the only way we should consider using that gun again is to read down a daily counter number, and let the people use that as a – we ran into a regular nightmare. Recording drinks used and drinks . . .

LOVELL — It was too much.

BORMAN — I think our water intake was adequate. And again I do not think this is something we need to be hounded about from the ground. A person seems to have enough problems without being reminded that he is drinking enough water; I do not see any reason at all to have this complicated bookkeeping system. The food intake was not a problem. We recorded the food. One thing we might note, the food was packed out of sequence. So when we got it out, we did not end up eating Day 1, meal A, B, C, Meal 2, A, B, C, but we did always eat Meal A first, Meal B the second meal of the day, and Meal C, the third meal of the day. Although, it might be day 1, meal A, day 13, meal B, and day 6, meal C.

65-HC-1001

LOVELL — Also, while we are talking about food, we better talk about the stowage of the food. It was right up to the maximum.

BORMAN — That is right.

LOVELL — And in fact it was a little bit overboard for 14 days. Rather tightly packed in. We really had to work to get it out.

BORMAN — But again, this is one of those things we were operating right up to the maximum capability of the spacecraft, and I guess we had to expect that, even though we did get it out. So I would not say it was unacceptable.

11.13 MISCELLANEOUS

BORMAN — Celestial and terrestrial observations, and significant observations or anomalies affecting other operational or experimental data.

LOVELL — Well, we got to see Mars just near the sun by looking at it just at sunset.

BORMAN — Mercury you mean. Mercury. We saw Mercury. And we made some other observations here for the scientists.

LOVELL — Zodiacal light came in loud and clear after we found out how to look for it.

BORMAN — Right.

LOVELL — We could not see the Gegenshein though we knew exactly where to look.

BORMAN — Okay. We also made several night passes counting meteors. One at 215 hours plus 23 minutes, 25 seconds, I made one night pass counting meteors for the entire 30 some minutes. And I saw 1 meteor under Taurus and Pleiades and it was below us, and it was short and white. The next night pass, ending at 217:17, I counted discrete flashes of lightning for the entire pass. I counted 206 discrete occurrences of lightning. This was with the spacecraft level in HORIZON SCAN mode, looking out one window with no yaw control. Now, we also saw individual meteors at different times in flight that are recorded on the tape recorder. But unfortunately we did not see this great shower that was supposed to come out of Gemini.

LOVELL — I saw two in a period of about 10 minutes out of Gemini. They headed below us, of course.

BORMAN — Alright, the meteors. Another significant observation that we made was a brilliant display of the Aurora. The Southern Aurora over Australia. And we have some pictures here.

LOVELL — Sketches.

BORMAN — Sketches of what it looked like, and we will cover this in the scientific debriefing.

LOVELL — And we measured the time it took for stars to occult to the air glow layer, and for Venus to go through.

BORMAN — I think this would be best to discuss when we talk to the scientists.

LOVELL — I think so.

BORMAN — Do you have anything else that was significant? As far as operational experimental data?

LOVELL — No.

BORMAN — We already mentioned the fact about the air glow changing with the Moon. Oh, I tell you one thing that is significant and we looked for it time and time again. This was the complete inability to observe stars in daylight.

LOVELL — Oh, yes.

BORMAN — I hope we put that one to bed, because we tried and tried and tried. We strained, we squinted, we looked at all angles.

LOVELL — Looked at all the angles.

BORMAN — Looked at all the angles, and we were never once able to observe a star in daytime. Now you can observe just at sunrise and just at sunset, but never in the daytime.

LOVELL — When that sun comes up those stars go.

BORMAN — That is right. And you get a black horizon – I mean a black sky above a blue horizon. I do not know what the reason for this is, but I will vouch for the fact that you can see it.

LOVELL — Right.

12.0 PREMISSION PLANNING

12.1 MISSION PLAN (TRAJECTORY)

LOVELL — Well, of course, it varies here, this is the last.

BORMAN — It changed. We had one all wired and written up for a 72 degree launch azimuth and this changed with the addition of the GT-6 mission. But I thought the people were very flexible, and adapted rapidly to the change.

LOVELL — Well, we are talking about the mission plan, and they are talking about the trajectory alone. There is no doubt about it that our mission, as originally set up for our fuel, was adequate. But, when we introduced GT-6 and the rendezvous mission, the amount of experimental work which we had to do also was not reduced at all. This compromised the results of our experimental work by having to use the fuel for the rendezvous. Although, I think the rendezvous was important, or that it was higher priority, I would like to have it put on the record, that the results we gathered from the rest of the flight were not as good as could be expected, because of the fact that we just did not have the fuel.

BORMAN — 110 pounds of gas went down the drain, too.

LOVELL — Yes.

12.2 FLIGHT PLAN

BORMAN — I thought the people did a great job there.

LOVELL — There is no doubt about it. The only way you can plan a mission of this length of time is real time flight planning. Call up the data you want for the day, it is a regular work day schedule. Call up what you want, and we will put it down and we will work at it; we will run it off that night.

12.3 SPACECRAFT CHANGES

BORMAN — I do not know of one major change in the spacecraft that we wanted that we did not get. GPO was very cooperative about everything. Of course, the big thing that we wanted, that we got after a hassle, was the ability to operate suits off. We planned this from the beginning with our first stowage review, and we finally made it by going the route of getting the suits that we could take off. And I think this contributed much to the satisfactory completion of the mission. Don't you? I have got personal notes here that were made during the flight. Every other page it says suits off was the only way to go, "I do not know how I stayed in the suit for six days," and so on. The suit I am sure has done more to increase the bugaboo of physical deterioration in space flight than any other single item. Far more than zero g.

12.4 MISSION RULES

BORMAN — They are routine now. We have no real arguments.

LOVELL — The only thing I can say about mission rules is the fact that they can be changed by the Flight Director to suit the situation. There is enough flexibility in them that allows the mission rules to meet the problem at hand.

BORMAN — Although, we did not have any problems that required us to change them.

LOVELL — No, we did not.

12.5 EXPERIMENTS

BORMAN — The only experiment that I thought was not well presented pre-mission was the laser. That was a sort of half baked preparation for quite a while, and at the last they brought the equipment down to Houston and it came out pretty good.

LOVELL — When we were first introduced to the equipment back at the stowage review up at St. Louis, we didn't know enough about the experiment to really analyze the equipment, to find out whether it would be adequate or not.

BORMAN — That is right. We should have picked up the reticule lighting on that.

LOVELL — Because we did not pick up the reticule lighting and we did not pick up the green filter.

BORMAN — No. The green filter looked all right when we used it on the ground. It is just the fact that we were not able to observe from far enough away to pick it up.

LOVELL — That is right.

BORMAN — The lighting contrast . . .

LOVELL — Also, experiments to the M-1 experiment was sort of a last minute glitch.

BORMAN — Yes.

LOVELL — We had a lot of compromise. Not compromise but a lot of failures.

BORMAN — Putting the M-1 into the ECS system you mean.

LOVELL — Although, I will have to admit that the whole thing worked out fine. The mechanics of the M-1 experiment did last 14 days and was absolutely no problem as far as operation of spacecraft or the ECS system.

BORMAN — Right.

LOVELL — It did complicate the pre-mission planning.

BORMAN — Every other experiment I thought was well presented. The experiments division, with Dick Moke helping out did a good job. I was very, very satisfied in other words, with the whole business.

13.0 MISSION CONTROL

BORMAN — Describe and discuss updating on the status of spacecraft and mission.

13.1
GO / NO GO's

BORMAN — Once a day operation, where we read out certain parameters in the spacecraft, reported them to the ground, and they have a GO or NO GO for the next area.

13.2
PLA AND
CLA UPDATES

BORMAN — They were extremely easy to handle. They came up in blocks of about 6 or 9. No problem, since we decided to use a rolling reentry in the event of contingency landing.

LOVELL — We had very little writing to do.

BORMAN — Very little writing. And I might add, based on that reentry, if I had to do it again, that is exactly what I'd want: a rolling reentry, if I did not have a load in that computer. Because I think it is very difficult to look out the window and observe a horizon during a reentry. At least it was during ours; especially during a night reentry.

13.3
CONSUMABLES

BORMAN — My goodness, we had 30 percent O_2 left when we jettisoned the adapter. 30 some percent FC O_2 and about 40 percent FC hydrogen. The OAMS were a little different situation. We cut that off at about 2 percent. I was a little disturbed on the real-time flight planning in the last couple of days when they were sending up for more experiments than they knew we had fuel to do. And telling us to sift them out. This sort of puts the onus of not doing it on the pilot. I guess in the long run it is the best way to do it, but they would call up and say, "well, here is a whole lot of updates that you can do if you have the fuel." Well, they knew darn well, we didn't have the fuel.

65-HC-1016

LOVELL — That is right.

BORMAN — In many cases it turned out to be weathered anyway.

LOVELL — And we had been briefed that there was a rather large error in the quantity readout on the quantity gauge, and fortunately for us it looked like it was in our favor.

BORMAN — We also had the Volkswagen tank which helped out. So we knew exactly where we were then.

LOVELL — But we should, in future flights, make sure that the spacecraft has enough fuel for adequate BEF alignment, and for at least a couple of revolutions with enough reserve to make sure that in case something goes wrong with the thruster, they could utilize more fuel to keep that alignment.

BORMAN — Right. Well, as it worked out, we came out all right, anyway. Because we just said we were not going to do anymore if we got to 5 percent to 6 percent. Then they came back and said that one time, I said, "Well, we're at 6 percent," and they said, "we thought you would go ahead and go to 5 percent." And this is an awful nebulous thinking.

LOVELL — Besides the gauge is hard to read to that accuracy.

BORMAN — That is right. Okay. We have already discussed, I think, the consumables on the fuel cells and the batteries. We turned off the squib batteries, about the 10th day. The fuel cell consumables were never a problem.

LOVELL — Main battery voltage was never really high, I thought. I have been led to believe that 22.5 volts was sort of a minimum voltage for a battery. They were up about 23 or 22.8 in the early part of the flight, and they were down to about 22.6, I guess, towards the end of the flight.

13.4 FLIGHT PLAN CHANGES

BORMAN — If there was a flight plan change, the only one that I know of, that we were not aware of, was the one of updating our perigee in the first burn. We thought we were going to burn it to 102. It came out we burned it to 120 and I'd like to know whether this was programmed, or whether we just burned too long. Other than that the flight plan went very well.

13.5 SYSTEMS

BORMAN — One of the items that I objected to a little bit in the flight, and it was the natural tendency with people on the ground; was the tendency, when we had a little systems malfunction, to explore it to the greatest depths without regard to the rest of the mission. For instance, when we had the failed thrusters and we were very, very low on fuel, they wanted me to put a 3 second direct burst through the thrusters; so they could get TM on what was happening. Well, this is fine, except when you do this, you are introducing the problem of the thrusters sticking, and losing all your fuel there. Or at least you are squirting out 3 seconds of valuable fuel, which is a heck of a big chunk in DIRECT. And if you were to induce ignition or even with three seconds of just venting fuel through the thruster you pick up a great rate, and then you have to stop; so you would use a tremendous amount of fuel this way. I will say when I refused to do it, though, they acceded, so it was all right. Another thing I didn't like, was this idea of blowing the OAMS squib. Remember? "This isn't in the flight plan" – "if you feel like doing it, I want you to blow the OAMS squib just to see if you can hear it." At this stage of the game we were depending upon the OAMS fuel for realigning the platform. I thought that by then we were in a stage of the mission that we were operational rather than interested in blowing something to see if we could hear it.

LOVELL — Especially for a night alignment.

BORMAN — Right. And I agree with you a thousand times it would not make any difference, but on the thousand and first time, it might have made a difference if we lost our OAMS, and I could not see any reason to do it; so we did not do it. Another item, that I did not like, came up as far as flight plan changes go. The request that was put in as part of the flight plan, and never was in the flight plan. That was to get a blood pressure over Guaymas after retrofire. As it turns out, we tried to do this, but we could not find the horizon and so we did not do it. I would strongly, as a matter of fact, I would not even consider it. After retrofire, as far as I am concerned, the blood pressures, and all the other non-operational equipment can go by the wayside.

LOVELL — Right.

BORMAN — Right then the important thing is to get on the ground.

LOVELL — Yeah, actually the retrofire time, the triple orbits before retrofire, when you are getting ready for reentry, should be exclusively devoted to that. We should be doing nothing else.

BORMAN — Right.

13.6 EXPERIMENTS REAL-TIME UPDATES

BORMAN — Experiments real-time updates: all we can say is that it was done fairly well.

14.0 TRAINING

14.1 GEMINI MISSION SIMULATOR

LOVELL — Since there were three crews that had been through these simulators quite extensively, and only one that required any more knowledge or any more operation of it, we decided to use the simulator at Houston as a systems trainer. Most of our basic training in the early phases of Gemini 7 was done on systems. We used the Houston trainer to gain knowledge of the systems. Towards the end we also used it for launch and reentry training.

BORMAN — We made no effort to keep it up to stowage configuration.

LOVELL — We made no effort to keep the Gemini Mission Simulator at Houston in any kind of a GT-7 stowage configuration. We just merely used the flying portions of it to get acquainted with systems and procedures. It was a procedures and systems trainer.

BORMAN — We got a lot of good reentries.

LOVELL — Well, that's procedures. We got a lot of good reentries, and we got a lot of good lift-offs with the trainer. One thing that helped us out quite a bit on procedures was the visual displays.

BORMAN — For stowage we kept the wooden mock-up in Houston up to our configuration. It came up better than I've ever seen it before. We did several exercises on the Gemini Crew Station Mock-up. We used that as distinct from the simulator for stowage.

LOVELL — The stowage mock-up was also used for experiments, to copy down updates, to put together equipment, to find out what electrical leads went where, and to practice using equipment inside the spacecraft. That way we didn't have to tie up the valuable Gemini Mission Simulator, that had a lot of electronics attached to it, with training that required only a simulation of space and environment. We acquired a pretty good knowledge of the systems with the simulator.

There's one area of procedures and systems training which can be improved, and that is the use of more correct procedures between ground and simulator. A lot of times we got into the simulator to do a reentry or launch, and we didn't get the parameters which you normally get from Mission Control. There have been five manned launches, with tapes on all of them, where the communications between the ground and the spacecraft are well documented. We ought to incorporate those in training. So you could copy down updates, you could get the 0.8, you could get systems failures and how the ground handles them, and things of this nature. I think we could simulate them a lot better now, especially in Houston.

BORMAN — I think Lynn Taegart was doing that.

LOVELL — Lynn was attempting to follow that procedure. We requested that they get some tapes from MCC from GT-5's launch, but by the time we left there, they hadn't yet arrived. Launch was very realistic.

BORMAN — I don't think that we flew a simulator with our roll program in it and it really didn't make any difference. I guess what we are saying about the Houston simulator is that we don't have to keep it right up to the final configuration. It's a basic trainer.

BORMAN — It helps a lot to get in there and get the work done, rather than not be able to get in it because it is being modified to bring it up to the latest configuration. As long as the math flow is proper, and the basic parts are there, you can almost leave it the same for every launch.

LOVELL — It should be updated as far as basic systems. We had the fuel cell panel put in ours; we wouldn't want batteries, because of a lot of systems training on fuel cells. That is important. We don't have to have every little item like on the Cape simulator for initial simulator training.

BORMAN — I was very pleased with the Cape simulator. We've had very good work out of it, and it was right up to the latest stowage configuration. I believe we were only shot down on the simulator once. It wasn't working perfectly every time we got in it, but at least we got valuable training out of it.

LOVELL — Right. One area that does need improving is the coordination between the simulators, the Cape and Houston.

BORMAN — You're talking about SIM NETS when we ran with Houston, which was almost a total waste of time, on our part.

BORMAN — That's right.

LOVELL — We wasted an awful lot of time just waiting around, because there wasn't the coordination between the two.

BORMAN — It was not the coordination, it was just the interface.

LOVELL — That's what I mean, the interface. They weren't connected properly. We did get a lot of bum dope; it wasn't the lack of training, it was the lack of proper training.

LOVELL — Station keeping with the booster occurred down here at the Cape only. I thought that was fairly realistic when it was working. I thought it helped some.

BORMAN — I thought it helped a lot.

LOVELL — Retrofire: procedure-wise was very good. The horizon, visual display, really is a big item. It made all the difference in the world between what we had before with no display and what we have now. As far as training goes, that is a very big item. The Houston Simulator ought to have a visual display as soon as possible. Retrofire reentries were all as programmed.

BORMAN — We had visual at Houston; we just did not have the targets. We had the stars at night.

LOVELL — We had the stars, but we had the occluding disc, and we did not have the horizon.

BORMAN — Reentry on the Mission Simulator was good, a very close approximation to what we flew in the spacecraft. I think that was very valuable. On GT-4, we had only flown about two or three reentries. We had to go up to St. Louis to fly these.

LOVELL — GT-3 and GT-4 were both that way. Essentially what we are saying about the Gemini Mission Simulator at Houston is that it is the basic trainer for systems and procedures. The one at the Cape is a fine mock-up for the final flight plan simulator. We go right through the numbers in the flight plan, and the SIM-NETS with the entire network.

14.2 DCPS (LAUNCH ABORT SIMULATOR)

BORMAN — We used it and it was very effective. It never worked closed loop as well as it was supposed to. We did not have any visual with it, but nevertheless the runs that we got there I thought were invaluable; we certainly could not do without them. The noise and the sensations seemed as close as you could get to them without running on centrifuge. It was a good program, the tapes were good.

LOVELL — I think it is a very necessary simulator. It was not working completely like we wanted it to work, but it was just being put into operation when we were at Houston. We did not have much of an opportunity to use it.

14.3 MAC ENGINEERING SIMULATOR

BORMAN — We ran a whole week at MAC, two days on reentries and three days on station keeping, and that was a very worthwhile week. It was concentrated effort, and the station keeping simulation was as close to what we saw as you could possibly get. It was just fantastic. It was really well done, and the reentries were also. We got a real good feel for the reentries. We had the people from FOD and the people from FCSD up there at the time. We understood not only the procedures for flying the reentries, but the why's, the how's, the limitations of the systems.

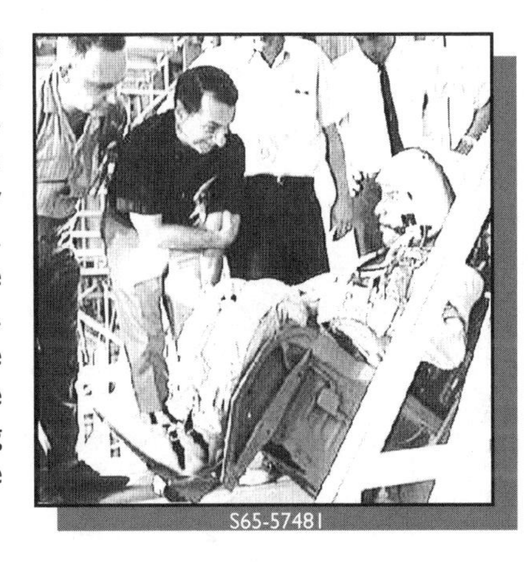
S65-57481

LOVELL — Well, that's where we dug out the procedures actually for the reentry technique.

BORMAN — That's right. That's where we developed the procedure of following the roll needle up to 3 G's. That was probably as good a week in training as we had the whole time.

14.4 TRANSLATION AND DOCKING TRAINER

LOVELL — In the little work that I did, it is representative of the actual case, if not more sensitive.

BORMAN — It's more sensitive. It is a more simple task to dock in space that it is in the Translation and Docking Trainer.

LOVELL — As long as it is more complicated there is no problem.

BORMAN — I was really surprised at that.

65-HC-1031

14.5 PLANETARIUM

LOVELL — The Planetarium is one which I have to admit I degraded for the last trip. I said I didn't want to go and we weren't going to go. Then we decided we'd better go, and I think right now it was well worth it. The last trip to the Planetarium was the best one.

BORMAN — We got more dope on the actual orbit. We had settled on our flight path, and they displayed it, and it was amazingly accurate.

LOVELL — We worked out our initial burn at the Planetarium, the stars we were going to burn on for our perigee adjust burn. I took a chart along that I made up at the

Planetarium to use for the zero, seven, and fourteen days in the celestial sphere. It worked out fine. When you go there, that's what you have to do. You have to take the azimuth that you are going to launch on, take all the charts that you are going to have; and run through that type of mission. To go there and just learn all the stars might be okay for basic training like we did several years ago. After you get assigned to a specific mission, you better start learning the stars you need to use for certain burns and things of this nature. That's a good place to do it. I thought perhaps, since we had the visual display, that we could eliminate the Planetarium because the visual display had more real feel for it, and it does. However, the visual displays in the simulators don't carry the magnitude of the stars that you can actually see.

BORMAN — And they are not flexible enough.

LOVELL — That's right. They are not flexible.

BORMAN — You can't change them.

LOVELL — So the Planetarium was helpful.

BORMAN — Spacecraft orientation. As Jim said, we studied the burns and made the first two burns on the stars. The Planetarium was very helpful for this. Remember, you were all set to find Corvus and then Spica and we had it all lined up before we ever launched.

14.6 SYSTEMS BRIEFINGS

BORMAN — We had one hundred and twenty-six hours and thirty minutes of system briefings and it's all well documented and scheduled. Another thing that was very helpful to us was Mike Brzezinski, the way he scheduled all the systems briefings and all the training. We didn't fool around with any of it. It was set up well, and went like clockwork. I think this was one of the real fine points of our training. Mike came down to the Cape a week early and the whole schedule was set up a week before we got here, and with very little change after that. He did an outstanding job, and as far as I am concerned, that's the only way to run it.

LOVELL — That's right. There's no sense running around doing it yourself when we have a nice, well run organization that can do this work for us.

BORMAN — As far as any figures or numbers on the time spent in the different phases of training, they are all available. If anybody wants them we have the final report, and we have a weekly report.

BORMAN — It was very helpful to us to just have to deal with one person, rather than dealing with the Planetarium people, the people at MAC, the people at the Cape. The only thing we did at all was contact Mike. We never contacted anyone else. We didn't go direct to anybody.

BORMAN — We might mention that the systems briefings were really of two types. We had preliminary systems briefings by our FCSD people at Houston. At McDonnell, during the SIMS flights and altitude chamber tests we filled in the dead time with systems briefings from the people at McDonnell. This was a good way to go also. For instance, while one crew was flying the MAC Rendezvous Simulator practicing station-keeping, the other crew was getting systems briefing. So, we didn't have any dead time. It worked out well.

BORMAN — The station-keeping on the booster was more difficult than it was on GT-6, primarily because the booster was venting, and tumbling and translating. Also because of the fact that we really didn't have a lot of time. We were having to use fuel in order to get set up before night, for the separation burn. However, once we got it squared away, there was no difficulty at all staying with the booster at a distance, I'd estimate from about sixty to one hundred and fifty feet. During the booster station keeping we did observe the ablative skirt on the engine. At one time it appeared that there were two points right at the edge of the skirt that might have been rolled in. It looked like maybe there were two approximately twelve inch sections that might have been rolled in. It may have been shadows cast off the booster. By and large, I would say that the engine looked very well.

LOVELL — To me the engine looked brand new. It had a gold cast to it. It looked perfectly good to me.

BORMAN — I did not notice any venting out of the roll nozzle, which is unusual. We thought the venting came out of the relief for the PSV valve on the side of the booster. We just discussed this with the Martin people. They were a little surprised to hear this. The venting we saw came out ninety degrees to the longitudinal axis of the booster.

LOVELL — It looked like it came out right at the edge of the tank.

BORMAN — And this is the Pressure Sequencing Valve drain. And the next one is, "If so, what was its condition?" It looked great. "Did you get pictures?" Yes, we got pictures with a 16 mm. We did not get pictures with the Hasselblad because we could not unstow the Hasselblad at this time.

14.7 FLIGHT EXPERIMENTS

BORMAN — Simulations: The one we used most frequently was the GMS for the D-4 / D-7 tracking.

LOVELL — We also used it for the sextant.

BORMAN — Although it wasn't a good utilization, we used it for the D-5.

LOVELL — Just for procedures. Most of our experiment simulation was done in the mock-up. The wooden mock-up.

BORMAN — Translation and Docking Trainer. We used the Translation and Docking Trainer for some tracking training with the laser. It was not all I had hoped it would be. Nevertheless it did give us training and convinced us it would be no problem in tracking with the laser. This was borne out by the flight that the main problem would be acquiring it. We picked up the fact in the docking trainer that the reticule is not visible at night. The Translation and Docking Trainer was helpful in station keeping. We didn't really do any training at McDonnell for the experiments.

LOVELL — We did mostly the station keeping and the reentry at McDonnell.

BORMAN — And you went to Ames.

LOVELL — Just for the sextant. The sextant briefing at Ames I thought could have been done at Houston. I got more information out of Bob Silva on the roof of the Cape here, than I did really out of Ames. The two simulators I used out there really didn't help me out at all. I think the simulator and a star field in the Gemini simulator plus work outside on the roof would be more valuable than Ames, which I passed on to Wally and Tom.

BORMAN — We had many briefings . . . Total time spent on experiments is two hundred hours and thirty minutes. A great deal of that time was briefing. I felt, that by the time we took off, we knew not only the procedures for every experiment, we knew the hows, the whys, and the wherefores. I thought we were adequately trained on every experiment. If I could comment on the one experiment that I thought was handled in a sort of, I won't say haphazard, but at least a rather free style, was the laser. We didn't get real proper training on that. I didn't get to look at the laser until later in the game. The ground equipment at Ascension never did come up, and I think that if we are really going to make this laser work, we are going to have to put more emphasis on the people who are running it. It seems to me that here we need some special procedure for training experiments. It seems to me that we went to some places where Mike was stuck with the position of trying to scrounge people and equipment or the individual experimenters were stuck with it. It seems like maybe Lilly should have been able to work with some section in our organization to get the training that we wanted and get it set up. The way it is now, for training on experiments, you almost have to depend on individual experimenters and a lot of the time they don't understand at all the problems of operating a spacecraft. Did you have that feeling ever, Mike?

FCSD REP — Just with the MSC-4. We did have a problem getting the equipment. All the other equipment seemed to come in well.

BORMAN — Yes, but we always depend upon the experimenters for the training. You know, like on the sextant. Maybe this is the way you have to continue to go.

FCSD REP — This is the way the program is set up, for the experimenters to actually do the briefing.

BORMAN — Yeah, and then the hardware training. That is the way it did work. That's probably the best way. The S-8 / D-13, we went to the trainer. I guess what we are really saying is that we should emphasize to the experimenter that they have a responsibility for providing training and for providing training hardware.

LOVELL — That is the big thing. The experimenters, or the experiments group, has to provide the training to get adequate results from their experiment. Otherwise they are not going to get adequate results.

BORMAN — That is right.

LOVELL — And the training equipment and the training periods have to come early enough in the program so that we work out any problems that evolve. For instance, a classic example was the laser when we ran into the reticule problem. We did not find out until too late in the game to change anything. We could not put a lighted reticule on the laser.

BORMAN — In all fairness, we ought to point out about the laser too, that it was severely handicapped when we changed the launch azimuth. Because initially they had not planned to work anywhere except at White Sands. Then with the change in launch azimuth, White Sands went down the tube pretty well, then they had to scrounge around and try to get to Hawaii and Ascension.

The experiment equipment, by and large, I thought was readily available on this flight. Thanks mainly to Lou Allen and the pressure he put on the people. Training equipment was pretty well available, early in the game.

14.8 SPACECRAFT SYSTEMS TESTS

BORMAN — We covered a lot fewer of the spacecraft systems tests than previous spacecraft had, based on our experience on Spacecraft 4. Let me see exactly what it was. Spacecraft tests, 169 hours and thirty minutes for the Prime Crew, and 193 hours and 30 minutes for the Backup Crew. I do not think there is any reason to cover things like Systems Assurance and so on in St. Louis. It is a waste of time. At St. Louis you should plan on covering the SIM flights, the Altitude Chamber and the Horizontal SEDR. And, down at the Cape, I thought even though we cut out six days of testing and we did not have a Wet Mock, I saw absolutely no impediment at all to our launch training. I do not think that any of it is necessary. I do not think we cut out one necessary thing.

LOVELL — I think that you could use your time more wisely in simulator training, in recovery training, and in training you are really going to use than in study of some of the systems assurance tests where you spend hours in the spacecraft just throwing switches. You reach a point there where you are not learning any more.

BORMAN — I think you should follow, at St. Louis, the SIM flight, the Altitude Chamber, and the Horizontal SEDR. And down here at the Cape, we want to do the SIM flight, the Joint Combined Systems test, the EIIV test, and finally of course, the SIM flight and Stowage Review.

LOVELL — That is something which we put in, and I think ought to be included in all . . . the stowage is one thing that changes constantly right after launch, and it ought to be put in just before launch. A week or so before launch to make sure everything is correct.

65-HC-943

14.9 EGRESS TRAINING

BORMAN — Briefing, Gulf Exercise, and Survival Gear. All went off well. We had it done on Spacecraft 4, so we only went off in Static Article 5. We did not use the Boiler Plate No. 201 and it worked out fine. We had the helicopter pickup.

LOVELL — I think the helicopter pickup was well worth it though, because it was exactly what you do on recovery. Might as well do it in practice.

BORMAN — And that is strange how that got thrown in there. That one time on Spacecraft 4, we just thought, well, it would be nice to come back by helicopter, rather than back by ship. And it worked out to be very valuable as a matter of fact. I think that is good training. I think that you should have the Gulf exercise. No question in my mind that you should have that.

LOVELL — That is about the most realistic type training you can possibly get.

BORMAN — We were well trained in the use of survival gear. Of course, I think that it is a very good idea. We had it laid out here in the crew quarters all the time we were down here, and we stopped in and took a look at it. We were thoroughly briefed on the ejection seat by NASA people. I thought we were well prepared for that.

14.10 PARACHUTE TRAINING

LOVELL — I think that all the parachute training that is required is launch off the island for a water landing. I think that is all the Parachute training you need, because that is most likely where you are going to land. You are going to land on land during an abort, so there would be all kinds of people to help you or you are not going to make it anyway. Guess you would be too close to the booster. I think the water landing training is very important, especially when you are using new equipment like our new suits. If we had the 4C suits again for this flight, I think that, since we had rotated so early from Gemini 4, that we could have eliminated that and not have any real problems.

BORMAN — I think that all the training should be conducted with training suits on. It does not make much sense to go out there in a swimming suit. And we might mention that we had one suit out there for our training. We switched in and out of a single suit, wet, or not wet, and got it all completed by one o'clock.

14.11 LAUNCH SIMULATIONS

14.12 REENTRY SIMULATIONS

14.13 SIMULATED NETWORK SIMULATIONS

14.14 NETWORK SIMULATIONS

BORMAN — Launch simulations, Reentry simulations, and Simulated Network Simulations, and Network Simulations, at the Cape down here, for us, were a total waste of time. We have already mentioned this earlier, but because of the fact that the simulator was not playing with the MCC. I would not say the launch simulations were a total waste of time. We did get some launches, but I am afraid that the time spent was not profitable.

LOVELL — We wasted an awful lot of time on that. It is not that they would not be profitable. I think that is really where you get the good training because you get . . .

BORMAN — I am not going to recommend eliminating it. I would recommend fixing it so that it plays properly.

LOVELL — That is right. Because you get the actual operating with the people that are going to be conducting the flight, get the communication procedures down, get the whole bit. Unfortunately the whole bit was not working.

BORMAN — That is right. I think this is recognized by all sides. I understand that Gemini 6 was much better after we left.

14.15 FLIGHT PLAN TRAINING

I really do not think you would call that training. It is sort of procedures that you go through, and I hope that the people that come behind realize how big a hand they can have in making the flight plan and how early they should get into the business, because there are so many people with their own little inputs. If the crew does not get in early and keep things under control, you will end up with an impossible situation.

Fortunately, I think, that the people that you have to work with now, Bill Tindall, and Barney Evans, are pretty good. One of the first things I would recommend to anyone to do, is to start talking to the Mission Planning people as soon as they get assigned to a flight. From then on, keep their fingers on the flight plan. I think that is reflected in the amount of time we spent on ours. 133 hours was spent in that training just on preparing and reviewing the flight plan. I think it paid off because we ended up with one that was reasonable and one that we could work with. I would not call it training though, as it was sort of doing flight planning.

15.0 CONCLUDING COMMENTS

BORMAN — What else do you have to say James?

LOVELL — We are back home, that proves the mission was a success.

BORMAN — There is one thing that I have to say I think the system we have set up here in FCSD now to handle these flights, this task force organization, is outstanding. I was very well pleased with the support we got from everybody.

LOVELL — I do not think we could have done it ourselves and having gone through it before when we did not have the organization set up this way, it sure made a difference.

BORMAN — And every part of the Center came through. It was very effective at the stowage mockups and the Design Review to have Kenny Kleinknecht or Chuck Matthews right there and to make a decision and then it stuck. That was very, very helpful from the very beginning there. The first day we went to that stowage review up there, we had the basic concept solved and we had the ECP's in to get the stowage the way we wanted it. It was very helpful. I do not think there was one thing that we really wanted in the spacecraft that GPO did not provide.

S66-15044

LOVELL — Everybody was very cooperative, I thought.

BORMAN — But that was another item that I strongly recommend the crews to do . . . is attend Management Meetings . . . particularly while the spacecraft is in St. Louis. You will find a lot of decisions are made and you can get in there and get your voice in. The people listen to you as long as you are not unreasonable, and you will end up making an awful lot of money in a very short time if you will get to the decision making. The important thing is keep a close tab on what goes on at the CCB. Jim Bilodeau is the best point of contact there. He kept us informed. For instance, let us say we wanted another stowage bag in the right footwell. Rather than just going through the back door and trying to get Carl Stone to make up a bag, we immediately submitted a requirement to the CCB. We found that time to react on this thing was amazingly short. You could get one in on a Thursday and it would be acted on by Monday. Then, after you had this clearance through the CCB, things went smoothly. I guess what I am saying is the system works and just plow in and use it. That is it

S65-62719

MANNED SPACECRAFT CENTER
HOUSTON, TEXAS
NATIONAL AERONAUTICS AND SPACE ADMINISTRATION

FACT SHEET 291-D
JANUARY 1966

GEMINI VII / GEMINI VI
LONG DURATION / RENDEZVOUS MISSIONS

The Gemini VII mission flown from December 4 to December 18, and the Gemini VI flight on December 15 and 16 completed a most successful year for the United States in the area of manned space flight. The only major step not yet accomplished in the Gemini Program is the successful docking of two space vehicles in space. The completion of the two December flights which combined the long duration (14 days) flight of VII with the rendezvous mission of VI accomplished two of the major goals of the program and permitted the United States to achieve a number of world records for manned space flight. They are:

- Longest manned space flight – Gemini VII, 330 hours, 35 minutes, 31 seconds; most revolutions, 206; most miles traveled, 5,129,400.
- First rendezvous of two manned maneuverable spacecraft – at times during the rendezvous the two spacecraft were within a foot of each other.
- First controlled reentry to a predetermined landing point – both Gemini VI and VII accomplished this feat.
- Most manned space flights in one year by one nation – five manned Gemini flights by the U.S. in 1965; also most men in space in one year, 10 (the United States has now had 16 astronauts with space experience).

The Gemini VII Spacecraft as photographed by Gemini VI during the fly-around maneuver following the rendezvous.

The two launches in December provided another first for the United States. The previous best "turn-around time" at the launch pad had been about two months. In order to launch VI in time to rendezvous with VII Gemini officials had to be reasonably certain that necessary repairs of superficial damage to Pad 19 could be made and that the Gemini VI spacecraft and launch vehicle final prelaunch preparations could be accomplished in a matter of days if the plan was to be followed. Work went so well that the first attempt at the rendezvous mission was made in slightly less than eight days after the VII launch. That launch attempt was terminated about 1.2 seconds after ignition when an electrical plug attached to the launch vehicle prematurely pulled free and caused an automatic shutdown. Three days after that attempt, Gemini VI was launched and successfully accomplished its rendezvous with Gemini VII about 185 miles above the earth.

GEMINI VII FLIGHT

On the morning of December 4, the countdown for the Gemini VII flight proceeded toward launch time smoothly and on schedule. The flight crew, Frank Borman, command pilot, and James Lovell, pilot, arose about 7 a.m. EST, and had a light snack consisting of orange juice, coffee, and toast. Several hours later they underwent a brief physical examination and were reported in excellent condition for the mission. Meanwhile, the backup crew – Command Pilot Edward White II and Pilot Michael Collins – who had been in the spacecraft and participating in checkout during the early part of the countdown, briefed Borman and Lovell on the status of the countdown before they entered the spacecraft. The VII crew had breakfast shortly before 10 a.m. It may have been that they wished to look at other faces during a meal or for some other reason, but at any rate that breakfast, which consisted of tenderloin steak, eggs, toast, jelly, orange juice, and coffee, was shared by 10 of their astronaut colleagues. They were John Young, Charles Conrad, Richard Gordon, Donald K. "Deke" Slayton, David Scott, Neil Armstrong, Virgil I. "Gus" Grissom, Alan Shepard, Walter Schirra, and Thomas Stafford. After breakfast the prime crew left the crew quarters on Merritt Island and traveled to the suit trailer at Launch Complex 16 where they donned the new light weight suits which had been developed for long duration missions. This suit weighs only 16 pounds, including an aviator's crash helmet which is worn under a soft helmet. The suit is so designed that it can be taken off during flight or can be worn in a partially doffed mode in which the gloves and boots are removed and the helmet is unzipped at the neck and rolled back to form a headrest. A total of 20 experiments were scheduled for the Gemini VII mission. These experiments will be covered in some detail in a later portion of this report.

A QUICK LOOK AT GEMINI VII / GEMINI VI

ACTIVITY	GEMINI VII (Frank Borman, Command Pilot) (James Lovell, Pilot)	GEMINI VI (Walter M. Schirra, Jr., Command Pilot) (Thomas Stafford, Pilot)
Liftoff (date and time)	December 4, 1965 2:30:03 p.m. EST	December 15, 1965 8:37:27 a.m. EST
Primary Mission Objectives	Long duration flight, Medical experiments, Target for Gemini VI	Rendezvous with Gemini VII
Results	Successful	Successful
Apogee (high point)	200 statute miles	194 statute miles
Perigee (low point)	101 statute miles	100 statute miles
Number of revolutions	206	16
Elapsed time of flight (hours, minutes, seconds)	330:35:31	26:01:40
Landing (date and time)	December 18, 1965 9:05:34 a.m. EST	December 16, 1965 10:39:07 a.m. EST

HIGHLIGHTS OF THE FLIGHT

Liftoff was at 2:30:03 p.m. EST. Following a nominal launch and orbital insertion, VII's crew performed a maneuver consisting of a 20-second propulsion burn which moved them away from the booster at about two feet per second. They moved to about 50 or 60 feet from the second stage of the booster, then performed a station keeping for about 20 minutes.

Figures on cutoff velocities and angles recorded 30 seconds after the sustainer engine cutoff show how close that portion of the flight went to the planned values. The slant-range from the Cape was 662 statute miles against a planned range of 665.8 statute miles. The velocity was 17,586.1 miles per hour, compared to a planned speed of 17,593.6 miles per hour. The perigee was 100.28 statute miles; the planned perigee was 100.05 statute miles. The apogee was 203.66 statute miles against a planned 210.45 statute miles. The planned inclination to the equator at that point was 28.87 degrees with the actual inclination 28.89 degrees.

Gemini VII, moments after liftoff on December 4, 1965, as it started its 14-day journey.

Later the crew performed a perigee adjusting maneuver designed to raise the perigee to 138 statute miles. By the sixth revolution the ground flight controllers had a very precise definition of the orbit – a perigee of 137.88 statute miles and an apogee of 200.21 statute miles.

From that time on, until their reentry after 206 revolutions, the crew was engaged largely in planned flight activities, in performing experiments, and in providing the target for the Gemini VI spacecraft on the 11th day of the flight.

During their second day of flight they photographed a tropical storm over the Indian Ocean. On the fifth day they made their major maneuver – one designed to circularize their orbit in order to offer the desired target for Gemini VI.

Prior to that maneuver, VII was in an orbit with a perigee of 146.74 statute miles and an apogee of 197.27 statute miles. The desired orbit following the maneuver was one with a perigee of 186.07 statute miles and an apogee of 187.56 statute miles. The actual orbit achieved had a perigee of 185.84 miles and an apogee of 187.33 miles.

DIFFICULTIES ENCOUNTERED

There were no major operational problems with the VII spacecraft, but, during the period of the mission, the following difficulties were experienced.

- The VII crew had difficulty with the reticule of a photometer developed for use in the star occultation experiment. During the next few days a number of actions recommended by the ground were attempted but they were not able to complete a "fix" and the experiment was canceled.

- About the end of the second day Command Pilot Frank Borman inadvertently pulled the EEG (electroencephalograph) electrodes used in the M-8 experiment loose as he reached back over his head. The crew attempted to replace the electrodes but this effort was not successful.

- On the sixth day trouble developed in one stack of one of the two fuel cell sections and this trouble continued to the end of the mission. On the 11th day another stack in the same fuel cell section failed. Sufficient electric power, however, was available at all times.

- The Gemini VII crew awakened during the seventh day of the flight and reported that the spacecraft was tumbling at an undesirable rate of 10 degrees per second. Engineers on the ground believed that this was caused by the water boiler exhaust, which imposes a tumbling movement on the spacecraft. This trouble was alleviated through a combination of by-passing the radiator to get rid of the water and by maneuvering the spacecraft.

- During the ninth day of the VII mission the onboard tape recorder which recorded non-medical telemetry data failed. The remainder of the mission was flown without this data, which was considered by Flight Director Christopher C. Kraft, Jr., as desirable but not critical to the success of the flight.

- On the 12th day the crew reported trouble with two attitude thrusters which resulted in a right yaw condition. This yaw was counteracted by use of the aft firing thrusters, costing only a minimum use of fuel.

The record breaking flight of Gemini VII came to an end on the morning of December 18 when the spacecraft reached the landing point in the Atlantic Ocean at 9:05:34 a.m. EST. Touchdown was within eight statute miles of the carrier WASP. Just 32 minutes after landing, the crew was landed on the deck of the WASP by the prime recovery helicopter.

The Gemini VII Crew, James Lovell, pilot, and Frank Borman, command pilot, leave the suit trailer on the way to Pad 19. They are wearing the specially-designed lightweight suits.

Dr. Charles A. Berry, Chief of Center Medical Programs, foreground, explains the operation of the Mission Control Center medical console to Dr. James Z. Appel, President of the American Medical Association.

William C. Schneider, Mission Director; Charles W. Mathews, Gemini Program Office Manager; and Christopher C. Kraft, Jr., Flight Director, (left to right), discuss progress of the Gemini VII flight.

GEMINI VI FLIGHT

The "third time proved the charm" for the Gemini VI spacecraft and its crew, Command Pilot Walter M. Schirra, Jr., and Pilot Thomas P. Stafford. VI had originally been scheduled to start its flight on October 25 and to rendezvous and dock with an Agena target vehicle. For the first time in manned space flight history countdowns were held concurrently for the Atlas-Agena and the Gemini launches. Both countdowns went without a hitch and the Agena was launched after Schirra and Stafford were in their spacecraft and counting down for their own scheduled launch about 95 minutes later.

Then contact was lost with the Agena several minutes after what seemed to be a perfect launch and it became apparent that that target did not go into orbit. When this became a certainty, as the stations girdling the world reported they were unable to acquire tracking signals, the flight of Gemini VI was scrubbed.

A display of equipment which is stowed in the Gemini VII spacecraft (below).

Several weeks later NASA officials announced that they would attempt to launch Gemini VI while Gemini VII was still in space and effect a rendezvous of the two spacecraft.

Dr. Charles A. Berry, Astronaut Elliot M. See, Jr., and George M. Low, Deputy Director of MSC, (left to right), are shown in the control room of Mission Control Center as they listen to communications from the spacecraft.

Two obviously happy space flight record holders, James Lovell, left, and Frank Borman are shown shortly after arrival on the carrier "Wasp."

Shown mid-deck on the "Wasp" is the antenna which picked up signals from onboard TV cameras during recovery operations of the Gemini VII / VI and transmitted live to the Early Bird satellite for relay to the United States.

Concentrated activity at the Cape following the VII launch made it possible to attempt the Gemini VI launch within eight days. Within 24 hours after VII was launched, the VI launch vehicle and spacecraft had been taken to the pad, erected, and mechanically mated. Less than one day later the spacecraft and the launch vehicle were electrically mated as the preparations continued.

During the next few days, the prime crew underwent continued flight training and a medical examination, a simulated flight was conducted with Mission Control Center in Houston tied in during that simulation (in effect supporting two missions simultaneously for the first time), and all work was geared toward the launch scheduled on December 12.

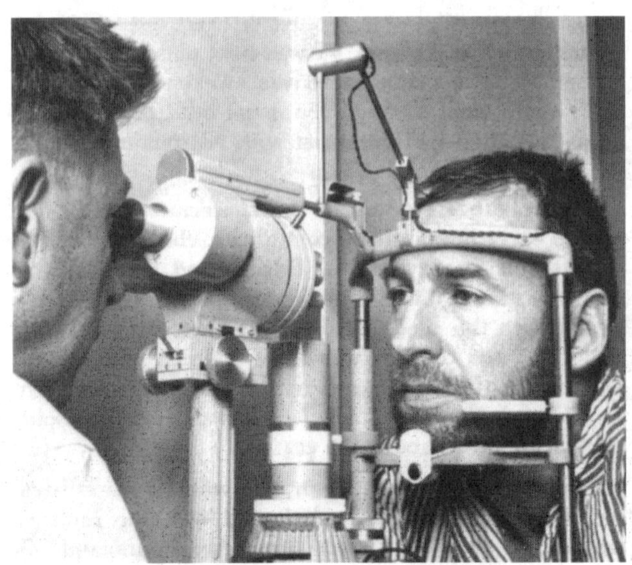

Lovell undergoes postflight medical checkup immediately after arriving aboard the aircraft carrier.

Early in the morning of that day, Schirra and Stafford were awakened, underwent a brief physical examination, then had the traditional preflight breakfast with Astronaut Gordon Cooper as their guest. They then proceeded to the suit trailer, and suited up for their mission.

In the meantime, the backup crew for VI – Command Pilot Virgil I. "Gus" Grissom, and Pilot John W. Young – was in the spacecraft during the early hours of the countdown.

The countdown continued uninterrupted toward the scheduled launch time of 9:54:06 a.m. EST. There was ignition at the right time but the engines were automatically shut down 1.2 seconds later. Schirra and Stafford correctly assessed the situation and determined it was safe to remain in the spacecraft.

A full Moon as seen from 185 miles above the earth, and photographed by the Gemini VII crew.

Immediately following that aborted mission steps were taken to determine the cause of the engine shutdown and an effort made to determine the amount of time required before another launch attempt could be made. It was determined that a small electrical plug in the tail of the launch vehicle had dropped out prematurely.

Later, as a result of data analysis and inspection of the launch vehicle, it was discovered that also a plastic dust cover had been obstructing the oxidizer inlet line of a gas generator. If the plug had not dropped out, this item would have precluded liftoff. It was pointed out that this dust cover, which is used as a contamination preventative, was in the launch vehicle on October 25, and would have prevented the launch on that day.

GEMINI VI IS LAUNCHED

The launch of Gemini VI was next planned for December 15. The prime crew was awakened at 4 a.m. EST; again went through the usual brief physical examination, then had breakfast with Astronaut Alan B. Shepard, Jr.; Gemini VI crew leader.

The backup crew was in the spacecraft again during the early hours of the countdown, checking on the various spacecraft systems. Schirra and Stafford entered the spacecraft 96 minutes prior to liftoff and proceeded with the checklist. Launch occurred at 8:37:27 a.m. EST, and Gemini VI was on its way, with a perfect launch, to its historic rendezvous in space with the Gemini VII spacecraft which was circling about 185 statute miles above the earth.

A series of maneuvers, beginning with VI's first pass over the United States, culminated with the rendezvous of the two spacecraft about five hours and 50 minutes after the liftoff of VI. The rendezvous was considered accomplished following a braking action by VI which brought the two spacecraft to within about 120 feet of each other.

At liftoff the VI spacecraft was about 1,380 statute miles behind VII. By the completion of the height-adjust maneuver over New Orleans during VI's first revolution, the distance between the two spacecraft was estimated at 730 miles.

Gemini VI starts toward its historic rendezvous with a target vehicle.

The Earth's horizon blends into the darkness of space to form an unusual background for this photo taken during the rendezvous.

Following two maneuvers during VI's second revolution, a phasing maneuver and a plane change maneuver, the distance between the two spacecraft had decreased to about 430 miles.

After the circularization maneuver in the third revolution, VI was about 15 miles below VII and they were about 190 miles apart. At the initiation of the terminal phase of the rendezvous about five hours and 15 minutes after VI's liftoff, VI was slightly below VII about 37 miles behind.

Activity in the Pad 19 Blockhouse prior to the launch of Gemini VI.

RENDEZVOUS ACCOMPLISHED

A wave of elation swept through the Mission Control Center in Houston when Stafford reported that they were within 120 feet of the VII spacecraft. Everyone in the Operations Room brought out a small American flag and fastened it to his console.

The VI spacecraft then closed in on VII and gradually got to the point where only one foot separated the two craft. VI then performed an in-plane fly-around maneuver around the other spacecraft, and later VII maneuvered beside VI. Gemini VI and VII flew in formation for about four revolutions, then Schirra performed another maneuver which put VI into an orbit with a higher apogee and lower perigee than VII. During the sleep period which followed for both crews, the distance between the two spacecraft varied from about 20 to about 40 miles. To obtain piloting experience in the rendezvous maneuver, all crew members took turns in the formation flying activities.

Several hours before retrofire, during a pass over the States, Schirra made the following report to Mission Control:

"This is Gemini VI. We have an object, looks like a satellite, going from north to south, up in a polar orbit. He's in a very low trajectory . . . looks like he may be going to reenter pretty soon. Stand by . . . it looks like he's trying to signal us." This transmission was immediately followed by "Jingle Bells," played by harmonica and bells. Thus, the spirit of the season was brought into the mission.

Speaking of that report, at the Gemini VII-VI news conference in Houston on December 30, Schirra said, in part, ". . . Our intent was not a prank. It was to relieve the tension . . . I think we convinced Chris and

Control Room scene immediately following rendezvous.

many of the people on the flight control team that we did, in fact, have an unidentified flying object there. And, I think the children of this country are happier for the fact that we might have seen something there."

The retrofire sequence took place about 700 miles northwest of Canton Island on December 16, and the VI spacecraft touched the water about 13 statute miles from the recovery vessel 30 minutes later. Schirra and Stafford preferred to remain in the spacecraft until it was brought on board the WASP, although both hatches were opened during part of the waiting period and the crew chatted with the frogmen who had attached the flotation collar around the spacecraft. The spacecraft was onboard the carrier one hour and two minutes after landing.

NASA got another bonus from the flight. The reentry and rendezvous section landed near the spacecraft and was retrieved by another team of swimmers. This is the first time this section, complete with the rendezvous radar equipment, has been retrieved. The main parachute was also retrieved.

NEWS CONFERENCES

Major news conferences were conducted following each of the launches, each of the reentries, and the December 12 attempt to launch VI.

NASA and Department of Defense participants at these news conferences praised the flight crews, the controllers, the ground support effort, the performance of the spacecraft and launch vehicles, the solid communications network performance, and the recovery forces.

They also cited the many achievements in the Gemini program particularly during 1965.

At the news conference, following the completion of the VI flight, Dr. George C. Mueller, NASA's associate administrator for Manned Space Flight, stated:

". . . This mission was the first time we had four people in space at the same time and the first time we had two spacecraft that flew in formation for several hours in space . . . it was the first time we have carried out a rendezvous in space. And, finally, it is the first time in the history of space flight that we carried out a successful controlled reentry, where the pilot controlled his landing point . . ."

In the news conference which followed the end of the VII flight, Dr. Robert R. Gilruth, director of the Manned Spacecraft Center, cited the achievements accomplished this year. He said:

". . . Since March of this year we have put 10 men into orbit and brought them back. We have accomplished the major part of the Gemini objectives at this point in the program. We have flown long duration flight . . . . We have seen extravehicular activity this year in Gemini, and we have seen rendezvous. We have seen demonstrated the controlled reentry technique that is so important to Apollo, and we have seen accomplished a whole raft of scientific experiments."

A Navy frogman gives the "OK" signal after attaching the flotation collar and checking the crew inside the Gemini VI spacecraft.

AWARDS CEREMONY

A special awards ceremony was held in the auditorium at Manned Spacecraft Center, December 30. James E. Webb, NASA administrator, made the presentations. Also present for the occasion was Congressman Olin Teague, chairman of the Manned Space Flight Subcommittee. They were introduced by George Low, deputy director of MSC.

The straps referred to early during the VII mission were photographed by the VI crew following their rendezvous.

Stafford and Schirra discuss the photo following their return.

Following opening remarks, Webb presented the NASA Distinguished Service Medal to MSC's Donald K. "Deke" Slayton, assistant director for Flight Crew Operations. The award was made "for his outstanding performance in directing NASA flight operations and for his leadership of the continuous and rapid adaptation of NASA's astronaut training activities to the experience gained from Mercury and Gemini flights . . ."

A large contingent of the "Wasp" crew observe the activity as the Gemini spacecraft is hoisted aboard.

The Gemini VI crew, Schirra, at right, and Stafford, leave their spacecraft. At the far right is Dr. Howard Minners, MSC physician

Webb awarded the NASA Exceptional Service Medal to the flight crew of Gemini VII, Command Pilot Frank Borman and Pilot James A. Lovell, Jr., and to the Gemini VI crew, Command Pilot Walter M. Schirra, Jr., and Pilot Thomas P. Stafford. In addition, Schirra was awarded the NASA Distinguished Service Medal for " . . . his courage and judgment in the face of great personal danger, his calm precise and immediate perception of the situation that confronted him and his accurate and critical decisions made possible the successful execution of the Gemini VI mission.

Others awarded the NASA Exceptional Service Medal were William C. Schneider, Office of Manned Space Flight deputy director for Mission Operations; and John T. Mengel, assistant director for Tracking and Data Systems Directorate, Goddard Space Flight Center.

Group Achievement Awards were accepted by G. Merritt Preston, Kennedy Space Center, deputy director for Launch Operations; Joseph M. Verlander, Martin Company, Gemini program director of the Canaveral Division; R. D. Hill, base manager, McDonnell Aircraft Corporation; Lt. Col. John G. Albert, chief of the Gemini Launch Vehicle Division, 6555th Aerospace Test Wing of the Air Force Systems Command; Lt. Col. Michael M. Kovach, chief of the Test Operations Division, Air Force Eastern Test Range; and John J. Williams, Kennedy Space Center assistant director for Spacecraft Operations.

A "fish-eye" view of the interior of the VII spacecraft.

ASTRONAUT NEWS CONFERENCE

The awards ceremony was followed by the news conference with both the Gemini VII and VI crews participating following brief opening statements by Webb and Low.

The VII crew described their mission first; then the VI crew talked about the rendezvous mission. Borman said that they found it very easy to stay with the second stage of their launch vehicle and attributed this ease both to their training simulations and to discussions they had with James McDivitt and Edward White who had attempted the stationkeeping exercise on their Gemini IV flight.

The VII crew was able to get pictures of a Polaris missile fired by the submarine Benjamin Franklin. Borman said, "This was quite an impressive sight. We saw the Polaris pop right out of the Atlantic at exactly the point it was supposed to be and we were able to track it very well."

Lovell told about the four orbit-adjust maneuvers that VII had performed. He pointed out that the first two were accomplished by "burning on stars." He said, "The ground crew called up a star that we should aim the spacecraft at. We timed the burn and just kept the attitude looking at the star. Both of these burns worked out well and both Frank and I feel that now if we do lose platforms or computers that we can make orbital-adjust maneuvers by visual observation only."

Discussing the successful liftoff of Gemini VII are, left to right, Lt. General Leighton I. Davis, USAF, DOD Manager of Manned Space Flight Support Operations; Dr. Robert R. Gilruth, MSC Director; and James C. Elms, Deputy Associate Administrator, Office of Manned Space Flight, NASA Headquarters.

SUITLESS COMFORT

Both men talked at length about the relative comfort of their suitless operation. Lovell removed his suit the second day. Later during the flight, he again donned his suit and Borman removed his. During the latter part of the mission both men were out of their suits.

The crew was asked whether they could have flown the 14 days if they had not taken the suits off. Borman said, "It is a subjective type question and I will say this – if we would have had the suits that had been used prior to our flight and not been able to remove them, I think our physiological effects would have been tremendous when we got out of the spacecraft. I am not certain that we would have been able to bounce back the way we did and it would have been a matter of survival rather than a matter of operating efficiently in space. Our firm recommendation is that on any long duration flight, the crew go suitless."

Lovell concurred, and added, "I think the motivation was there – we could have gone for 14 days in just about any space suit. But the matter of how well you operate in space is much improved with the lightweight suit."

The Gemini VII flight was the first in which both pilots were scheduled to sleep at the same time. That schedule went into effect after the first day. The VII crew had music piped up to them periodically and both men had a book along which they read in the evenings for a while before going to sleep. Appropriately, Borman chose Mark Twain's "Roughing It." Lovell chose "Drums Along the Mohawk," by James Fenimore Cooper.

STAGING DESCRIBED

In telling about that mission, Schirra said that at staging there was a very large orange flame with a brownish-black edging to it that enveloped the spacecraft and that they flew through it. He described it as leaving a smoke residue layer on the windows.

Speaking of the same event, Stafford said that the flash at staging was really fantastic. He said "It started from the back corner. It was a yellowish-orange phenomena . . . at the edge of it was a brownish-black . . . and in just one second we flew right through the whole phenomena . . .

Schirra then discussed the maneuvers they made in order to effect the rendezvous. He said, "The burns that we made to adjust our orbit, to change our catch-up rate, to raise our perigee, were almost classic, they were so pure. Our updates were exactly as we had hoped they would be. The data came to us very nicely. The greatest delight I suspect both Tom and I had was when we made our tweak burn.

TWEAK BURN OUTLINED

"To describe what this tweak burn is: we were making burns on the order of forty to fifty feet per second to adjust the various phases of the orbit. At the termination of the major velocity burn, we would have residuals, or little bits of velocity to clean out in all three axes. This might amount to three-tenths of a foot per second. On one occasion, one was about seven-tenths of a foot per second. Then we made minute little burns to bring these velocities toward zero. After all this wrestling, final phase adjust or height adjust came up from the ground, 'you have an .8-foot per second burn to make.'"

"As we progressed through the various maneuvers to effect a point in space which we describe as the NSR – this is the co-elliptic burn – we realized that we were working with a very professional team who had fed us all of the good information we needed, that we could do no mission as rendezvous is described without such information. This is part of the mission where the ground has to put us in the right hole. They put us through the insertion window for the right orbit. The timing was exquisite. They put us into the circularization burn exactly the same way. Basically, it's almost like going into orbit all over again. I think this is probably the mission where man's ingenuity in developing computers has really come to bear. There were many computer complexes involved – computers running in real time could tell us where a star was that Frank and Jim used for pointing for their burns, computers that told them what velocity to burn; and then we had our on-board computer that we finally had to rely upon."

ACCURATE DATA FLOW CITED

Stafford said, "It was really a great experience to see the interface of the ground and the onboard capabilities work, and the way the data would flow. Needless to say, when we first turned on the radar set at the programmed time, Wally and I were both a little anxious to see the green lock-on light go steady. It didn't so right away I said, 'Well, here's what I've been trained for.' In other words, to go through all the backup modes and make lots of computations. We did get a real positive lock-on in excess of 230 miles. From 230 miles down to 60 feet, that radar set performed beautifully. It told us exactly when we went from 120 feet down to 60 feet; it told us when we went from 235 miles down to 234.5 miles."

Schirra said they did see spacecraft VII with their light from about 22 to 23 miles away. From that point on they could see spacecraft VII all the way into intercept. The light used was called the docking light, it was installed on VII and on VI to illuminate another object in space, such as the Agena or the booster in Frank's case. He added, "Somebody from another place said when you come within three miles, you've rendezvoused. If anybody thinks they've pulled a rendezvous off at three miles, have fun! This is when we started doing our work. I don't think rendezvous is over until you are stopped, completely stopped with no relative motion between the two vehicles at a range of approximately 120 feet, that's rendezvous. From there on, it's station keeping, that's when you can go back and play the game of driving a car or driving an airplane or pushing a skateboard, it's about that simple. As we progressed into Gemini VII, using the docking light as our target, we finally came into a range of about 700 feet and that is when I lost respect for the lights that are illuminating us now. This sunlight on VII was absolutely the brightest thing I have ever seen in my life; my eyes hurt; it was a klieg-like carbon arc lamp completely saturating my eyeballs."

GEMINI VII EXPERIMENTS

A total of 20 medical, technological, and engineering experiments were planned for the VII mission. Because of the nature of the experiments and the fact that many of them are conducted as one of a series of similar flight experiments, only a preliminary evaluation of the results is available at this time. In most cases detailed evaluations and conclusions will not be completed until after all the data for each experiment has been analyzed.

- Two of the VII experiments concerned Celestial and Space Object Radiometry. Purpose of these experiments was to obtain spectral irradiance information about terrestrial features on celestial objects. A total of 37 measurements was made. Of a possible 32 subjects to be radiometrically measured, 29 were accomplished. Early study of data indicate that the experiment was very successful.

- Objectives of the Star Occultation Navigation experiment were to determine the usefulness of star occultation measurements for space navigation and to determine a horizon density profile to update atmospheric models for horizon-based measurement systems. This experiment failed to provide any data due to an instrument malfunction. The failure analysis identified the exact source of the problem and established a means for correction.

- The Simple Navigation experiment was designed to provide data on observable phenomena of space flight which could be used to solve navigational problems by using optical data to calculate an actual space position by manual techniques. A sextant was used in this experiment. During the flight, the VII crew made a total of 37 star-to-horizon, six star-to-star, five planet- or star-to-moon, and eight star-zero measurements. The crew noted difficulty in acquiring and identifying stars when they were near the bright moon. Additional work with this experiment is anticipated on later flights in order to obtain additional data.

- The Inflight Sleep Analysis experiment was intended to accomplish three objectives: (1) to assess the pilot's depth of sleep during orbital flight in order to optimize work-rest cycles and to correlate sleep depth with pilot performance; (2) to assess the effects of the weightless state on the wave patterns of the normal electroencephalograph (EEG); and (3) to assess states of diminished alertness and relate these to the general inflight command pilot performance. This experiment was only run with the command pilot who had four EEG electrodes attached to his scalp. The electrodes were inadvertently pulled loose during the 56th hour of the flight. A "quick look" examination of the tapes has shown that usable data were obtained during the duration of the experiment. Analysis of the data is continuing.

- Another medical experiment conducted during the VII flight concerned the Human Otolith (inner ear) Function. Objectives were to measure any change in otolith activity associated with the flight, and, particularly to measure any change that might result from prolonged weightlessness. An onboard vision tester was used in this experiment. Both the command pilot and the pilot were given a series of

Space Sextant used on Gemini VII simple navigation experiment.

preflight and postflight tests in addition to the daily inflight tests conducted. Preliminary evaluation of data indicates that each pilot rendered accurate and consistent visual estimations of the horizontal position of the spacecraft. In addition, postflight debriefing reports indicate that neither pilot experienced any disorientation during the flight, despite the fact that they underwent a series of intentional head shaking movements carried out with their eyes closed.

VARIED EXPERIMENTS

- The objective of the Proton-Electron Spectrometer experiment was to measure proton and electron intensities outside the spacecraft and the dose being received by the crew during the mission. The integrated radiation dose was monitored with operational film-badge packages located in the crew underwear. A check of the flight log indicated the experiment was operated as planned through the first eight days of the mission. At the end of that period, however, a failure in the spacecraft telemetry recorder prevented obtaining any data except those which could be transmitted in real-time to ground stations along the orbital track. Preliminary examination indicates that the experiment resulted in good electron data. An erratic response in the equipment indicated an intermittent failure in obtaining proton information. Dose estimates will be calculated after all data from the spectrometer are reduced.

- An Optical Communications experiment was also conducted on VII. Objectives were to evaluate an optical communications system, to evaluate the astronaut as a pointing element, and to probe the atmosphere using an optical coherent radiator outside the atmosphere. Equipment used consisted of a laser transmitter on the spacecraft and a flashing beacon and equipment capable of collecting and demodulating coded optical signals at ground sites located at White Sands Missile Range, New Mexico; Kauai Island, Hawaii; and Ascension Island in the South Atlantic. Unfavorable cloud conditions hampered this experiment throughout most of the Gemini VII mission and all but four of the scheduled attempts were canceled. The ground beacon was observed only twice in these four attempts. Although none of the objectives were attained, the major system parts were proven.

MORE MEDICAL EXPERIMENTS

- The objective of the Bone Demineralization experiment was to investigate the occurrence and degree of any bone demineralization resulting from prolonged space flight. X-rays were taken of the heel bone and the terminal bone of the little finger of the left hand of both crew members prior to the flight and another series of X-rays was started immediately following recovery. No conclusions as to the results of the experiment were available at the time of the publication of this report.

- The Inflight Exerciser experiment was conducted to assess cardiovascular reflex activity in response to a known physical workload and to determine the capacity of man to perform such work under prolonged spaceflight conditions. The inflight exerciser consisted of a pair of rubber elastic cords attached to a handle at one end and to a nylon foot strap on the other end. Seventy pounds of pressure were required to pull the exerciser to its full extension – 12 inches. Two exercise periods were scheduled each day for each crewman. During these periods, the pilots pulled the handle once a second for 30 seconds. Blood pressure measurements were made before and after each exercise period. This experiment was classified as a success. The crew demonstrated their ability to perform physical work through 14 days of flight.

- Another medical experiment, the Inflight Phonocardiogram, was designed to investigate the functional cardiac status of the crew during prolonged space flight. A small microphone was attached to the pilot's chest wall at the cardiac apex during the sensor application portion of the suiting up procedure prior to the flight. Heart sounds were recorded on an inboard biomedical recorder.

- A Bioassays Body Fluids experiment was conducted to determine the physiological effects of space flight. To accomplish this, body fluids obtained during preflight, inflight, and postflight periods are analyzed for electrolytes, hormones, proteins, and other organic constituents.

VISUAL ACUITY TESTED

- Two other experiments investigated Visual Acuity and Astronaut Visibility. The objectives of these experiments were to measure the visual acuity of the flight crew members before, during, and after long duration space flight in order to determine the effects of a prolonged period in space environment; and to test the use of basic visual acuity data combined with (1) measured optical properties of ground objects and their natural lighting, (2) the atmosphere, and (3) the spacecraft window to predict the limits of naked-eye visual capability in discriminating small objects on the surface of the earth during daylight. Equipment used consisted of an inflight vision tester, an inflight photometer used to monitor the spacecraft window, and a test pattern at a ground observation site.

The ground observation site was located 40 miles north of Laredo, Texas. Eight 2,000-feet squares of plowed, graded and raked soil were arranged in a four by two matrix. White rectangles of Styrofoam-coated wallboard were laid out in each square in north-south, east-west, and diagonal positions and the pilots recorded their identifications. The rectangle orientations were changed between passes and the size adjusted in accordance with anticipated slant range, solar elevation, and the visual performance of the flight crew on preceding passes.

Prior to the flight, both Borman and Lovell completed 12 sessions in a laboratory training van during which they became experienced in the techniques to be used and established physiological baselines descriptive of their individual visual performances. The objectives of these experiments were successfully achieved. Preliminary evaluation of the results obtained indicated that the visual performance of both men was not degraded during the 14-day mission.

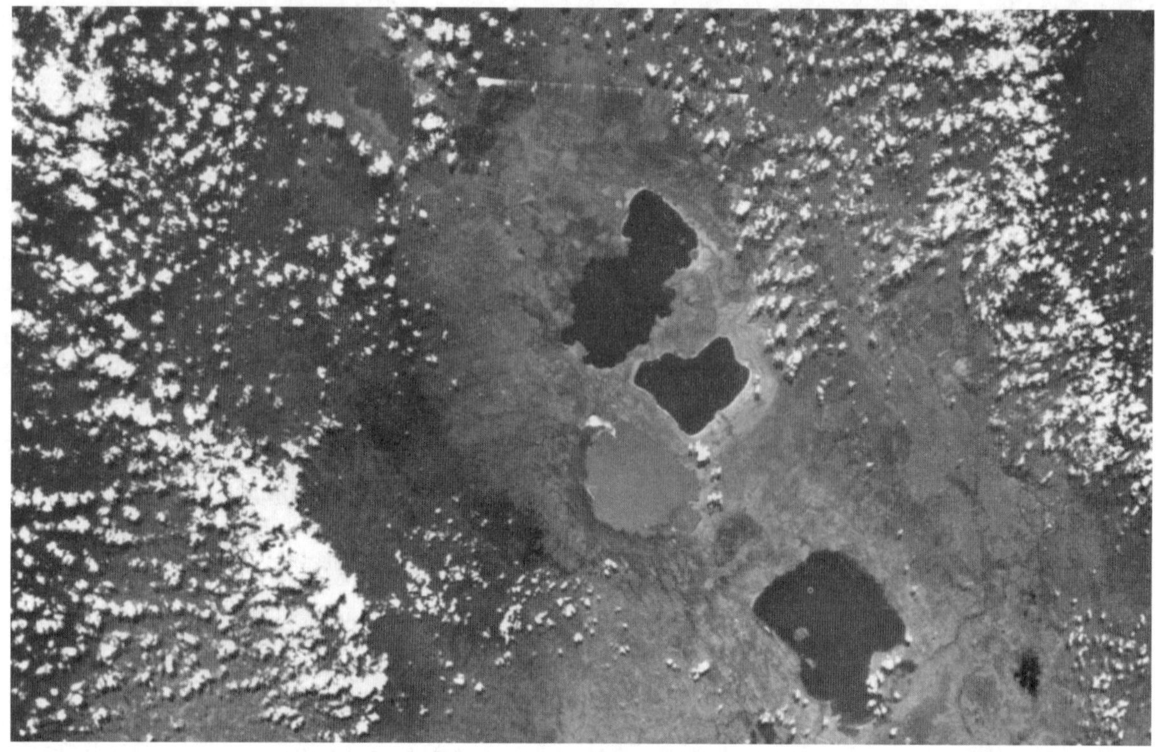

A photo taken from space of an area in Ethiopia, south of Addis Ababa.

A photo of the Aden protectorate in Arabia.

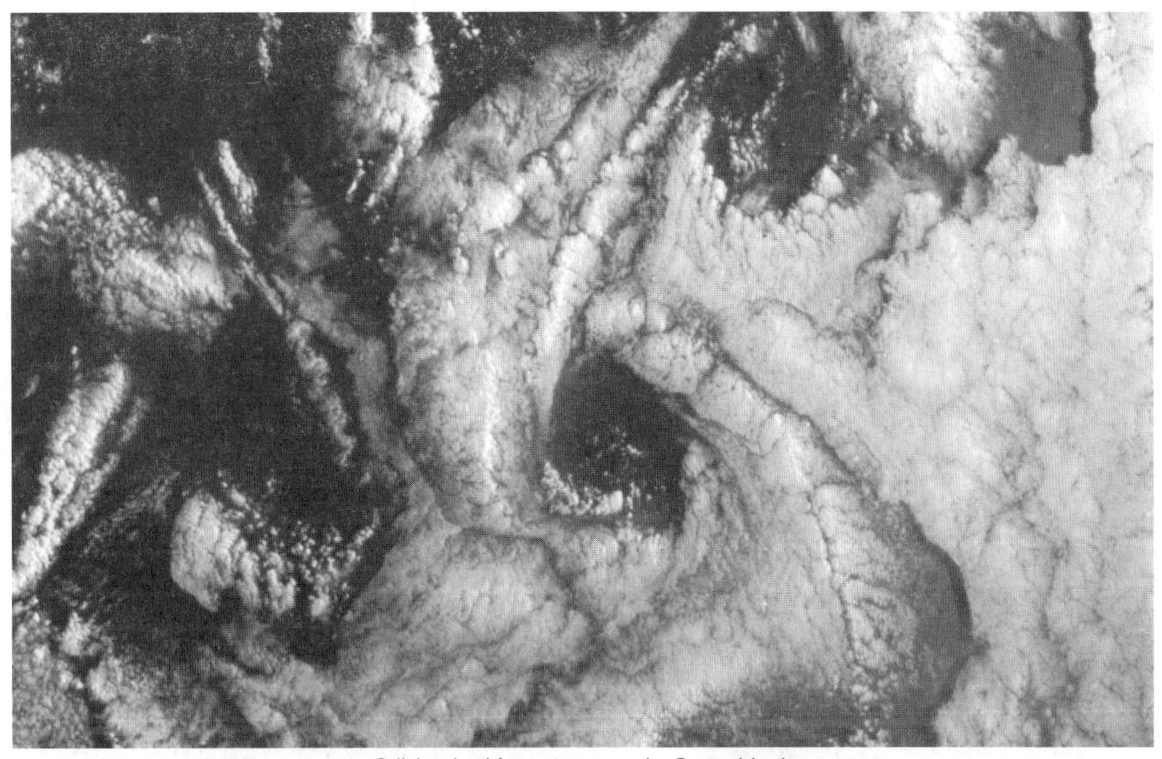

Cellular cloud formations over the Canary Islands.

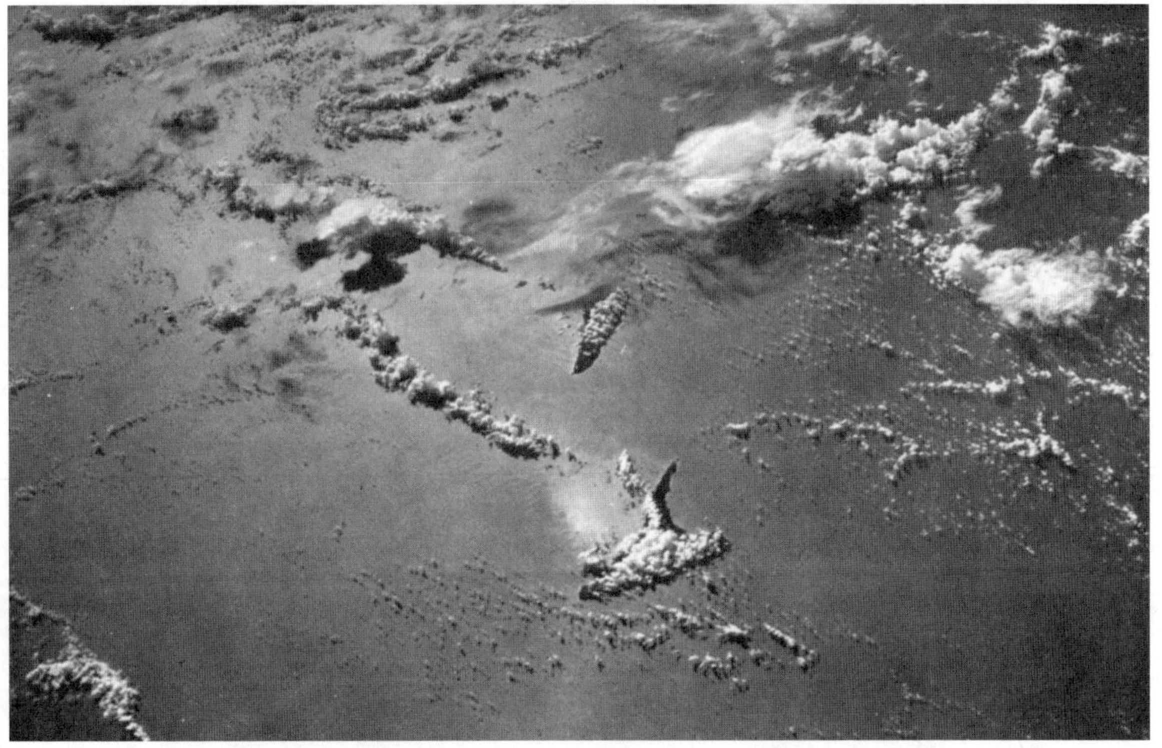

Clouds over the Maritius area throw unusual shadows over the surface of the water.

LANDMARK, SYNOPTIC PHOTOGRAPHY

- The Landmark Contrast Measurement experiment was designed to measure visual contrast of landmarks against their surroundings. These data were to have been compared to calculated values of landmark contrast in order to determine the relative visibility of terrestrial landmarks as seen from outside the atmosphere. These landmarks, when visible, are expected to provide a source of data for the onboard Apollo guidance and navigation system. Experimental equipment for this experiment was the photometer scheduled for use on the Star Occultation experiment. Because of a malfunction of that photometer, no data were obtained. Postflight technical debriefing comments by the crew, however, gave indication that the experiment is operationally feasible.

Testing Equipment used during the Gemini VII mission in performing the visual acuity experiment.

- Color infra-red film was used successfully for the first time in a manned flight in the Synoptic Terrain Photography experiment. The objective of that experiment was to obtain high-quality small-scale photographs of terrain and ocean areas for geological, geographical and oceanographic research. Pictures of selected areas in Mexico, Africa, Australia, South America and preselected ocean areas were particularly desired. A large number of photographs of the selected areas were obtained. Several continuous sequences suitable for stereoscopic study were obtained over portions of Africa and Mexico, as well as a number of oceanographic pictures showing the Bahama Islands which should be useful in studying changes in bottom topography caused by Hurricane Betsy.

- The objective of the Synoptic Weather Photography experiment was to obtain color photographs of a variety of cloud systems for meteorological studies. These high resolution pictures will supplement those taken on the Gemini IV and V missions and provide a comparison of the Earth's weather in equatorial regions during different seasons. The results of this experiment have been classed as a success. About 190 weather photographs were obtained.

THE PILOTS

Frank Borman

The VII Command Pilot, Frank Borman, was born in Gary, Indiana, March 14, 1928, but spent most of his early years in Tucson, Arizona. He is five feet, 10 inches tall; weighs 163 pounds; and has blonde hair and blue eyes.

Borman was graduated from the United States Military Academy in 1950, chose an Air Force career, and received his pilot training at Williams Air Force Base, California. From 1951 to 1956 he served with fighter squadrons in the United States and in the Philippines and was an instructor at the Air Force Fighter Weapons School.

In 1957 Borman received a master of science degree in aeronautical engineering from the California Institute of Technology, and from 1957 to 1960 he served as an instructor of thermodynamics and fluid mechanics at the Military Academy at West Point.

He graduated from the USAF Aerospace Research Pilots School and later served there as an instructor. He was selected as an astronaut in September 1962. Borman has logged more than 4,400 hours flying time, including more than 3,600 hours in jet aircraft.

He is married to the former Susan Bugbee of Tucson, Arizona. The Bormans have two sons – Frederick, 14; and Edwin, 11. His parents, Mr. and Mrs. Edwin Borman, reside at Phoenix, Arizona.

The Gemini VII crew, Command Pilot Frank Borman and Pilot James A. Lovell, Jr.

James Lovell

James A. Lovell, Jr., who served as pilot for the VII mission, was born in Cleveland, Ohio, March 25, 1928. He is six-feet tall, weighs 165 pounds, and has blonde hair and blue eyes.

He is a graduate of the United States Naval Academy. Following his flight training, he served in a variety of assignments, including a three-year tour as a test pilot at the Naval Air Test Center, Patuxent River, Maryland. His duties there included service as a program manager for the F4H Weapon System Evaluation.

Lovell was graduated from the Aviation Safety School of the University of Southern California, and he served as flight instructor and safety officer with Fighter Squadron 101 at the Naval Air Station, Oceana, Virginia. He was selected as an astronaut in September, 1962. He has amassed more than 3,000 hours flying time, with more than 2,000 hours of that in jet aircraft.

Lovell is married to the former Marilyn Gerlach of Milwaukee, Wisconsin. They have three children – Barbara Lyn, 12; James A., 10; and Susan Kay, 7. His parents, Mr. and Mrs. James A. Lovell, Sr., live at Edgewater Beach, Florida.

Walter Schirra

Walter M. Schirra, Jr., command pilot of Gemini VI, was born in Hackensack, New Jersey, March 12, 1923. He is five feet, 10 inches tall; weighs 170 pounds, and has brown hair and brown eyes.

Schirra was graduated from the United States Naval Academy in 1945, and received his flight training at Pensacola Naval Air Station, Florida. As an exchange pilot with the Air Force, he flew 90 combat missions in F-84E aircraft in Korea. He downed one MIG and was credited with another probable. Schirra was awarded the Distinguished Flying Cross and two Air Medals for his service in Korea.

He took part in the development of the Sidewinder missile at the Naval Ordnance Training Station, China Lake, California. Schirra was also project pilot for the F7U3 Cutlass and instructor pilot for the Cutlass and the FJ3 Fury.

He attended the Naval Air Safety Officer School at the University of Southern California, and completed test pilot training at the Naval Air Test Center, Patuxent River, Maryland. He was later assigned at Patuxent in suitability development work on the F4H.

Schirra has logged more than 3,800 hours flying time, including more than 2,700 hours in jet aircraft. He was one of the seven Mercury astronauts named in April 1959 and was pilot of the six-orbit Mercury-Atlas 8 flight on October 3, 1962.

Schirra is married to the former Josephine Fraser of Seattle, Washington. They have two children – Walter M. III, 15; and Suzanne, 8. His parents, Mr. and Mrs. Walter M. Schirra, Sr., live in San Diego, California.

Thomas Stafford

The Gemini VI Pilot, Thomas P. Stafford, was born in Weatherford, Oklahoma, September 17, 1930. He is six feet tall, weighs 175 pounds, and has black hair and blue eyes.

Stafford is a graduate of the United States Naval Academy. Following his graduation, he was commissioned in the Air Force. After completing his flight training he flew fighter interceptor aircraft in the United States and Germany and later attended the Air Force Experimental Flight Test School at Edwards Air Force Base, California.

He served as Chief of the Performance Branch, USAF Aerospace Research Pilot School at Edwards. In this capacity he was responsible for supervision and administration of the flying curriculum for student test pilots. He also participated in and directed the writing of flight test manuals for use by the staff and students. He became an astronaut in September 1962.

The Gemini VI crew, Command Pilot Walter M. Schirra and Pilot Thomas P. Stafford.

Stafford has logged more than 4,300 hours flying time, including more than 3,600 hours in jet aircraft. He is married to the former Faye L. Shoemaker of Weatherford. They have two daughters – Dianne, 11; and Karin, 8. His mother, Mrs. Mary E. Stafford, resides in Weatherford.

UNITED STATES SPACE FLIGHT LOG

Mercury-Redstone 3	Shepard	May 5, 1961	00:15:22	00:15:22
Mercury-Redstone 4	Grissom	July 21, 1961	00:15:37	00:30:59
Mercury-Atlas 6	Glenn	Feb. 20, 1962	04:55:23	05:26:22
Mercury-Atlas 7	Carpenter	May 24, 1962	04:56:05	10:22:27
Mercury-Atlas 8	Schirra	Oct. 3, 1962	09:13:11	19:35:38
Mercury-Atlas 9	Cooper	May 15-16, 1963	34:19:49	53:55:27
Gemini-Titan III	Grissom-Young	Mar. 23, 1965	04:53:00	63:41:27
Gemini-Titan IV	McDivitt-White	June 3-7, 1965	97:56:11	259:33:49
Gemini-Titan V	Cooper-Conrad	Aug. 21-29, 1965	190:55:14	641:24:17
Gemini-Titan VII	Borman-Lovell	Dec. 4-18, 1965	330:35:31	1302:35:19
Gemini-Titan VI	Schirra-Stafford	Dec.15-16, 1965	26:01:40	1354:38:39

GEMINI VII VOICE COMMUNICATIONS

(EXCEPTED FROM THE AIR-TO-GROUND, GROUND-TO-AIR AND ON-BOARD TRANSCRIPTION)

RENDEZVOUS

TANANARIVE

263:28:18	P6	. . . burning are being run pretty well . . .
263:28:21	*C7*	*Roger.*
263:28:23	*P6*	*. . . in about 30 degrees, Frank.*
263:28:25	C6	Okay.
263:28:35	P6	20 seconds to go 2 minutes.
263:28:39	C7	I'll never make it.
263:28:40	C6	. . .
263:28:42	P6	. . . Frank.
263:28:43	C7	Roger. 3 minutes.
263:28:59	C7	You say you're pitching out at 3 degrees now?
263:29:01	P6	About 3 degrees.
263:29:03	C7	Right . . .
263:29:09	P6	How does it burn?
263:29:12	C7	We can't see . . . I hope they're working.
263:29:22	P7	We've got a very dim . . . light.
263:29:25	P6	Can't see any thrusting lights.
263:29:27	P6	Negative ...
263:29:50	P6	. . . 3 minutes . . .
263:29:53	P6	5, 4, 3, 2, 1
263:29:59	P6	MARK.
263:30:00	P6	3 minutes.
263:30:01	C7	Roger. Made it.
263:30:03	P6	. . .
263:30:44	P6	A real dim light up there . . .
263:30:47	P7	We'll blink it a couple times.
263:30:58	C7	. . .
263:31:01	C7	It's off now.
263:31:03	P6	Okay . . .
263:31:05	C7	It's coming on now.
263:31:07	P6	Roger . . .
263:31:15	C7	And we don't have any ACQ lights . . . zero.
263:31:40	P6	We've about 35 to go.
263:31:43	P7	Roger.
263:31:51	P6	. . .
263:31:59	C7	. . . yet.
263:32:00	P6	. . .
263:32:05	C7	You want me to do that or do you want me to stay where I am? I better stay where I am I guess.
263:32:08	P6	. . .
		COASTAL SENTRY QUEBEC
263:50:04	P7	. . . retrofire.
263:50:09	CC	How are you doing there, Jim?
263:50:12	P7	. . . a little bit too low.
263:50:15	CC	Roger.
263:50:51	P7	. . .
263:51:03	CC	Very good.
263:51:06	CC	How are you fixed now?
263:51:08	P7	Well, I'm reading 10 degrees.
263:51:11	CC	. . .
263:52:25	P7	110 degrees.
263:52:27	CC	I now have Goddard cut off.
263:52:59	P7	Very good.
263:54:45	C6	. . . 120 degrees.
263:54:49	C7	. . .
263:54:54	C6	1.7 miles.

263:54:56 C7 Roger.
263:55:42 C6 . . . I have you . . .
263:55:44 C7 Very good.
263:55:51 C6 125 degrees and 1.3 miles.
264:04:24 C7 You got a lot of stuff all around the back end of you. Must be the . . . off from the . . . test . . .
264:04:46 C7 You want me to stay stationary now, Wally? We're almost face up.
264:05:09 C7 This is VII.
264:05:11 C6 Go ahead, Frank.
264:05:12 C7 I'm going to go ahead and put it on Inertial – Neutral here and stay right on the horizon, if that's what Precious wants.
264:05:18 C6 Great . . .
264:05:19 C7 Say again.
264:05:21 C6 That will be fine, Frank.
264:05:22 C7 Okay.
264:07:01 C6 You guys really are showing a . . . of droop on those wires hanging there.
264:07:07 C7 Stop . . . it on me.
264:07:10 C7 Where are they hanging from?
264:07:13 C6 Well, Frank, it looks like it comes out at the separation between – it might be the Fiberglass. It's approximately – oh – l0 to 15 feet long.
264:07:22 C7 The separation came from the booster, right?
264:07:24 C6 Affirmative.
264:07:25 C7 That's exactly where you have one, too. It really belted around there when you were firing your thrusters.
264:07:32 C6 Looks like about 8 or 9 feet long and double wire.
264:07:35 C7 Right.
264:07:36 C6 We're going to take a picture of it.
264:07:53 CC VII, Hawaii.
264:07:55 C7 Go ahead.
264:07:56 CC I've got a short flight plan update if you'd like to copy it.
264:07:59 C7 Stand by one.
264:08:12 C7 Go ahead, Hawaii.
264:08:14 CC Okay. D-4 / D-7: 265:43:00 –
264:08:19 CC – Sequence 427 –
264:08:34 CC I say again. Sequence 427; Mode 03; Spacecraft VI, one minute of recorder. D-4 / D-7: 265:44:00; Sequence 427; Mode 01; Spacecraft VI, 2 minutes. Note: to be performed during Gemini VI tape playback at Hawaii.
264:09:28 CC Did you copy that all right?
264:09:32 C7 Roger.
264:10:56 CC VI and VII, Hawaii. We'll be standing by if you have anything for us.
264:11:00 C7 Roger.
264:11:01 P6 There just seems to be a little traffic up here, that's all.
264:11:04 CC Call a policeman.
264:11:08 P6 It was pretty trying during the terminal run. As we looked out, we could see the two Gemini stars casting off to the right of Gemini VI – correction, VII. They were all in a line.
264:11:21 CC Roger.
264:11:49 CC VII, Hawaii. Would you turn your adapter C-Band off?
264:11:52 C7 Roger.
264:11:55 C6 It's in the COMMAND position.
264:11:59 CC Okay, . . . that will do it.
264:13:46 C7 We've lost you, VI.
264:13:49 C6 You've lost sight of me, sir?
264:13:50 C7 Roger.
264:13:54 C6 I must affirm a few little things.
264:13:56 C7 Okay.
264:14:03 C6 Keep to your left now.
264:14:04 C6 This is the night light, is it?
264:14:07 C7 Say again.
264:14:08 C6 This doesn't act like a night light, or is that an experiment light?
264:14:11 C7 No. They should both be off.
264:15:43 P7 The flag or the letters are visible. Looks like they're seared as much at launch as they are when you come back from reentry.

264:15:51	P6	Jim, noticed your blue field is practically burned off.
264:15:55	P7	Right.
264:19:15	C7	Elliot, Gemini VII.
264:19:17	CC	Go ahead.
264:19:18	C7	Roger. This fuel cell is dropping down again. You want us to take the Platform off the line?
264:19:25	CC	Stand by.
264:19:29		. . .
264:19:32	C7	We'll keep the Platform mode if you want.
264:19:35	C7	. . . Wally.
264:19:37	C6	Say again.
264:19:38	C7	I put my automatic on the . . .
264:19:40	C6	Okay.
264:19:50	CC	Gemini VII, Houston. We'd like to leave the Platform on and take Stack 2C off the line at this time.
264:20:00	C7	Stand by, sir. Okay, here you go.
264:20:16	C7	Stack 2C is off the line.
264:20:18	CC	Roger. What was your current on it before you took it off?
264:20:22	C7	About 5½ amps. It deteriorated from 10 to 5½.
264:20:27	CC	Roger, VII.
264:20:49	CC	Gemini VII and VI, would you continue with the description of your stationkeeping?
264:21:03	C7	Right now, VI is about 10 feet above and to the left of VII. We're just flying nose to nose, approximately 15 feet apart.
264:21:13	CC	Roger.
264:21:15	P7	We can very clearly see the horizon scanners up . . .
264:21:21	CC	Roger, Jim. Gemini VII, are you able to see in the windows of VI very easily, and vice versa?
264:21:36	C7	Roger. VII here.
264:22:40	P7	Wally, I can see your lateral thrusters are firing out about 40 feet from what we – visual, from what we can see.
264:22:48	C6	How are they doing?
264:22:50	P7	We can't tell now. We're in too close. But when you were doing your Braking Maneuver we could see them . . . quite a bit out.
264:23:09	P7	I'll come back in, so we can get nose-to-nose a little bit.
264:23:25	P6	Looks like those wires guillotined off the booster and not at the spacecraft . . .
264:23:31	C7	Yes, you have the same thing, only it's in back of yours.
264:23:34	C6	The wire bundles, of course.
264:24:09	P7	Houston, on this Braking Maneuver our lateral thrusters fire quite aways out.
264:24:14	CC	Roger, VII.
		TEXAS
264:24:58	CC	Gemini VII, Houston.
264:25:00	C7	Go ahead, Houston.
264:25:02	CC	We plan to put 2C back on the line at the RKV.
264:25:07	C7	Okay.
264:25:08	CC	Approximately 20 minutes off the line.
264:25:11	C7	Roger.
264:25:32	CC	Gemini VII, Houston. Could you give us a readout on your stack currents?
264:25:36	C7	Stand by.
264:25:38	P7	Roger. Houston. 1A is 2½ amps; 1B, 11 amps; 1C, 9½ amps; 2A, 8½ amps; 2B, 7½ amps; 2-Charlie about to zero.
264:25:52	CC	Roger, Jim.
264:25:54	P7	The configure voltage on 2-Charlie is reading 31.2.
264:25:58	CC	Roger. Copy. 31.2.
264:26:05	CC	Looks very good, VII. You might keep an eye on 2-Charlie voltage and see if you can see it jump up like it did yesterday.
264:26:12	P7	Roger.
264:26:17	CC	Gemini VII, would you switch your adapter C-Band to CONTINUOUS?
264:26:24	P7	Adapter C-Band CONTINUOUS.
264:26:25	C7	. . .
264:26:26	P6	Yes. What's confusing?
264:26:39	C7	What did you say, Wally?
264:26:40	C6	You guys sure have big deals.
264:26:42	CC	Ha! ha! ha!
264:26:45	C6	If you're in white, you're in style.

264:26:47 CC Right.
264:36:50 P7 Well, let's see about 264:36 now. Formation with GT-VI. Taking pictures and passing the time of day.
264:38:59 C7 12 o'clock tomorrow morning?
264:39:01 . . . after tomorrow morning.
264:39:05 . . . easier ones on Saturday.
264:39:08 I'll pass.
264:39:12 C6 How's the food supply holding out?
264:39:14 C7 Oh, it's in good shape.
264:39:16 P7 It's holding out, but it's the same thing day after day.
264:39:26 . . .

ROSE KNOT VICTOR

264:39:40 CC Gemini VII, RKV.
264:39:41 C7 Go ahead, RKV. Gemini VII.
264:39:43 CC Roger. We'd like you to bring Stack 2C back on the line.
264:39:46 C7 Coming on.
264:39:58 C7 Stack on.
264:39:59 CC Roger.
264:40:00 CC Did you get an open-circuit voltage before you put it on, please?
264:40:04 P7 Looks like it's in the same group as it was when you took it off open.
264:40:07 CC Roger. What was your open-circuit voltage before you turned it back on?
264:40:11 P1 31.5 volts.
264:40:12 CC Say again.
264:40:14 P7 31.5 volts.
264:40:15 CC Roger.
264:40:19 CC What are you reading now? That's 2C current.
264:40:25 P7 2C current is reading 6 amps and closed-circuit voltage is a little below 25.
264:40:31 CC Roger.
264:40:37 CC We're reading 5.5 on 2C on the ground.
264:40:38 C7 . . . RKV.
264:40:44 CC Say again, Gemini VII.
264:40:46 C7 I say we have company tonight.
264:40:48 CC You sure do. We'd like to know how you can see the ACQ lights on VI.
264:40:54 C7 He doesn't have ACQ lights.
264:41:00 P7 They have a dock light control.
264:41:03 CC Did they see the dock lights?
264:41:04 C6 . . . one point off. Right, yes.
264:41:08 P7 We picked you up, Wally, when the sun reflected off your adapter.
264:41:11 C6 Yes.
264:43:51 CC Gemini VII, we'd like you to take 2-Charlie off the the line at 265.
264:43:58 P7 Roger. 2-Charlie off the line at 265:00.
264:44:03 CC Roger.
264:44:05 C7 How long do you want us to leave it off the line?
264:44:08 CC We'll give it a check at CSQ.
264:44:11 C7 Okay.
264:45:36 C7 Hey, Wa!ly.
264:45:37 C6 Go ahead.
264:45:39 C7 When we get around to the next stateside, let us try it for about 5 minutes, will you? I'm after – the fuel is bounded up a little bit.
264:45:41 P6 Sure.
264:45:45 C7 Okay.
264:45:49 C7 I'm going to Platform for the nighttime pass, if you want to back off just a little bit.
264:45:53 C6 When I get in there you're going to leave the Platform up?
264:45:56 P7 Yes, we're going to leave it up.
264:45:58 C6 Okay.
264:47:00 C6 That will help you there, Frank, or . . . back a little more?
264:47:01 C7 That's okay. I . . . in the plat mode now.
264:47:04 C6 Roger.
264:47:07 P7 . . . on the line right now.
264:47:09 C6 Roger.
264:47:16 C7 This gage is bouncing right around 15 percent if you look at it right. . . . to the left . . .
264:47:23 P7 Oh, good. I think we better pitch over.

TANANARIVE

265:02:22	P6	Spacecraft . . .
265:02:24	C7	Say again.
265:02:26	P6	. . . just talking about.
265:02:29	C7	You're not close.
265:02:30	P6	. . . ring you up.
265:02:31	C7	That's your dump.
265:02:33	P6	Oh, okay.
265:02:40	P7	Can you see Frank's beard, Wally?
265:02:42	C6	Yes, I see yours better right now.
265:02:47	P6	Did you wipe your mouth, Jim, after you ate?
265:02:52	P6	You must have just wiped your mouth, Jim. Did you?
265:02:56	P6	Yes. Right.
265:02:58	P6	How's the visibility through these windows? They're pretty bad from this side.
265:03:02	P7	Yes. It's pretty bad. We noticed . . . that some see through the windows. They are pretty bad on this side?
265:03:14	P7	Through your window, they're right on top of us.
265:04:30	C6	The forest fires really kick them out, don't they, Jim?
265:04:32	F7	Right. They've been there all the time, Wally.
265:04:34	C6	Yes. That fire down there to your left is an oil fire, I think.
265:04:40	P7	Oh, you see one there from the air?
265:04:43	P6	Right. Maybe that was down to your left. It's . . . night.
265:04:48	P7	Okay.
265:04:55	P6	. . . any way you do it.
265:04:59	P7	Between the black marks.
265:05:04	P7	With your light on, Wally, I can just see the flame of the front of the nozzle.
265:05:10	C6	Roger.
265:05:40	CC	Gemini VII, Houston. Can you confirm you turned 2-Charlie open-circuit again?
265:05:47	P7	This is VII. Roger. 2-Charlie is open.
265:05:50	CC	Roger. What does the open-circuit voltage look like?
265:05:53	P7	Looks like about 31.2 volts.
265:05:57	CC	31.6. Roger.
265:05:59	P7	31.2.
265:06:00	CC	31.2.
265:06:06	P7	Drinking water, right, Wally?
265:06:09	C6	Roger.
265:06:11	P7	. . . voltmeter very much to look at.
265:06:16	C7	. . . rough day.
265:06:17	C6	Ha! ha!
265:06:21	C6	Don't let him kid you. I'm just a . . .
265:06:24	P7	Roger. . . .
265:06:46	C6	. . . Merry Christmas and get it over with.
265:06:50	P7	Yes. I still wish we had a pot of fuel here.
265:06:59	C6	You should work up a . . .
265:07:01	P7	Right.
265:07:31	C6	. . .
265:07:34	P7	That's right.
265:08:07	P6	Back up just a little bit.
265:08:09	P7	There doesn't seem to be any shape of a . . . just a glob comes out.
265:08:14	P6	. . . ball of fire attitude . . . light out all you can see is the initial flame.
265:08:24	P7	. . . for awhile.
265:08:31	P7	As you approached us in the rendezvous we could see the fire way out to about 40 feet.
265:08:37	P6	Very good.
265:08:43	P7	Now we didn't even know you were there, Wally.
265:08:45	C6	Right here.
265:08:46	P7	Yes.
265:08:47	C6	Roger. . . .
265:08:49	C6	Ha! ha!
265:08:52	C6	Now I can see you a little more . . . taking off on the . . .
265:08:58	P7	My . . . off.
265:09:05	P6	No. I guess it's because I get so many reflections on the window.
265:09:14	P7	Roger. We saw that.

265:09:16 P6	. . . sunlight.
265:09:19 P7	Roger. I could hit you with it.
265:09:22	. . .
265:09:24	Very good. . . . on the next flight.
	COASTAL SENTRY QUEBEC
265:25:15 CC	Gemini VII, CSQ.
265:25:17 C7	Go ahead, CSQ. Gemini VII.
265:25:19 CC	Roger. Will you give us another reading on 2-Charlie open-circuit voltage.
265:25:23 P7	Roger. 2-Charlie open-circuit voltage is 31.8 volts.
265:25:29 CC	Roger.
265:25:34 CC	Would you place 2-Charlie back on the line?
265:25:38 C7	2-Charlie coming back on the line.
265:26:21 CC	Gemini VII, CSQ. We'll probably leave 2-Charlie on at least until Hawaii, and have a look at it there.
265:26:28 C7	Roger. I read: 2-Charlie, remain on to Hawaii and take a look.
265:26:32 CC	Roger.
265:29:25 CC	I don't . . .
265:29:28 C7	. . . your flight plan, Hawaii.
265:29:31 C7	7 hours 22 minutes . . . for a vent.
265:29:34 CC	Check 22.
265:29:35 C7	Roger.
265:29:42 C6	We'll switch to your experiment, then we'll fly around, if that's all right with you.
265:29:47 C7	Let us have about 5 minutes . . . trying to move around in this mode.
265:29:52 C6	Oh, fine.
265:29:56 C7	In fact, how about that right now?
265:29:57 C6	Sure, my Maneuver switch has been off for the last 15 minutes.
265:30:00 C7	Okay, we'll just coast around here a little bit and see what it looks like.
265:30:03 C6	Very good.
265:30:08 C7	This experiment will start in about 13 minutes, this morning.
265:30:11 C6	Roger . . . Flat node and Node 2.
265:31:17 C7	I know . . . with you.
265:31:23 C6	It's looking pretty good.
265:31:25 C7	Yes, those heaters work fine.
265:31:39 C7	Yes, you're right.
265:31:41 C6	. . .
265:31:43 C7	That's what I mean – there are times when it looks like there's nothing there.
265:31:46 C6	Yes, a couple of white flakes, and that's about it.
265:31:49 C7	Yes.
	HAWAII
265:41:23 C6	We'll be dumping in about 2½ minutes. Radar contact.
265:41:24 CC	Gemini VII, Hawaii.
265:42:11 CC	Gemini VII, Hawaii CAP COM.
265:42:15 P6	Go ahead, Hawaii. VII here.
265:42:17 CC	How you doing?
265:42:18 P7	It's great! Really outstanding!
265:42:20 CC	Okay. Give me a readout on 2-Charlie, please.
265:42:24 P7	Roger. 2-Charlie closed circuit is reading 24.8.
265:42:27 CC	Roger.
265:42:30 P7	. . . amp . . . 5½.
265:42:33 CC	Roger, very good.
265:42:35 C7	Hawaii, this is VII, here. We're going to fire on the platform, unless Flight has any objections, and get back on the regular schedule and leave the spacecraft in the HORIZON SCAN.
265:42:49 CC	They say: "Have at it."
265:42:50 C7	Thank you.
265:43:34 CC	Gemini VII, Hawaii CAP COM.
265:43:36 P7	This is VII. Go.
265:43:37 CC	If you haven't started powering-down yet, they want to schedule a purge, Jim.
265:43:40 P7	Okay, we have not started yet, so let's purge.
265:43:43 CC	Okay, hold up a minute.
265:43:50 CC	Okay, we're ready for your purge. Have at it.
265:43:52 P7	Purging Section 1.
265:43:53 CC	Roger.

265:43:56	CC	One little bubble.
265:44:03	C7	Boy, those windows are really bad. I got a good look at your window, Wally. It's really coated.
265:44:08	C6	Yes.
265:44:10	C6	We were lucky.
265:44:20	CC	All right.
265:44:24	C6	. . . that's a real big problem there.
265:44:26	C7	Wally, can you tell if there's any purge at all?
265:44:29	C7	Say again.
265:44:30	C7	Could you tell if we're purging at all?
265:44:34	C6	I see some white flakes, bubble, white things come off. Not actually bubbles, but . . .
265:44:49	C7	Could you move down a little bit so you're more in line with us?
265:44:52	C6	Moving down.
265:45:29	CC	VII, put your Quantity Read switch to ECS O_2 position.
265:45:33	CC	Okay, just leave it there, please.
265:45:35	P7	Roger.
265:46:26	CC	VII, Hawaii. Go to FUEL CELL O_2 position in Quantity Read.
265:46:29	C7	Done.
265:46:30	CC	Okay, just hold it there, and we'd like it for about 15 minutes for the power-up and hold off on your power-down until 266 plus 01.
265:46:41	C7	266 plus 01. Roger.
265:46:43	CC	Okay.
265:47:05	CC	Quantity Read to the FUEL CELL H_2 position, please, VII.
265:47:09	C7	We are.
265:47:10	CC	Okay.
265:47:41	CC	VII, Hawaii. Quantity Read switched off.
265:47:44	CC	Okay, thank you.
265:47:46	C7	Okay. We've been using the O_2. Do you want it off tonight?
265:47:51	CC	Say again.
265:47:52	C7	We've been leaving the Fuel Cell O_2 on because of that stuck transducer. You want me to turn if off tonight, Bill?
265:47:56	CC	Okay. They'll give you a good briefing before you go to bed. If you don't mind, you can leave it in the O_2 position now.
265:48:02	C7	Okay, I'll leave it off.
265:48:45	C7	And Hawaii, for your information, VII is terminating here with about 11 percent.
265:48:50	CC	Thank you, VII.
265:49:01	P7	Hawaii, purge complete.
265:49:03	CC	Roger. Got the whole thing. Thank you very much.
265:49:43	CC	VII, Hawaii. We've got nothing further. We'll be standing by if you need anything.
265:49:46	C7	Thank you, Hawaii.
265:49:49	C7	2-Charlie in the second section looks sick again tonight.
265:49:53	CC	Well, we'll see what's going to happen. Just hang in there.
265:50:06	C7	You going over the top, Wally?
265:50:08	C6	That's right.
265:50:09	C7	Okay.
265:50:10	CC	Standing by one.
265:50:11	C7	The adapter about the experiment, top of Jim's head, is all clay, like an acid or something came out on it.
265:50:19	P7	It's a liquid Freon or Neon or something from the cold IR experiment.
265:50:24	C6	Oh, is that what it is?
265:50:25	P7	Yes.
265:50:55	C7	How does it look from back there?
265:50:57	C6	Looks good.
266:15:24		. . . check over . . .
266:15:28	C7	. . . load is looking?
266:15:31	C6	We're looking right at your vehicle now.